PENGUIN BOOKS

THE ORIGINS OF VIRTUE

Matt Ridley's books include the bestselling *Genome: The Autobiography of a Species in 23 Chapters* and *The Red Queen: Sex and the Evolution of Human Nature.* He received his D.Phil. in zoology from Oxford University, and previously worked as science editor, Washington correspondent, and American editor for the *Economist.* He lives in Newcastle, England.

Praise for *The Origins of Virtue*

"[Matt Ridley] manages to combine a scholarly approach with a great dash and wit, which puts him well ahead of the field; stimulating and great fun."
—Max Wilkinson, *Financial Times*

"In an era in which biological science is challenging traditional ethics, he has raised the debate to a new level of seriousness and importance." —John Cornwell, *The Times*

"The book is extremely well written with the sort of anecdotal detail and wit that make for lively reading even when the most abstract topics are being treated."
—Frans B.M. de Waal, *Nature*

"If my *Selfish Gene* were to have a volume two devoted to humans, *The Origins of Virtue* is pretty much what I think it ought to look like." —Richard Dawkins

Matt Ridley

THE ORIGINS OF VIRTUE

Human Instincts and the Evolution of Cooperation

PENGUIN BOOKS
Published by the Penguin Group
Penguin Group (USA) Inc., 375 Hudson Street, New York, New York 10014, U.S.A.
Penguin Group (Canada), 90 Eglinton Avenue East, Suite 700, Toronto,
Ontario, Canada M4P 2Y3 (a division of Pearson Penguin Canada Inc.)
Penguin Books Ltd, 80 Strand, London WC2R 0RL, England
Penguin Ireland, 25 St Stephen's Green, Dublin 2, Ireland (a division of Penguin Books Ltd)
Penguin Group (Australia), 250 Camberwell Road, Camberwell,
Victoria 3124, Australia (a division of Pearson Australia Group Pty Ltd)
Penguin Books India Pvt Ltd, 11 Community Centre, Panchsheel Park, New Delhi – 110 017, India
Penguin Group (NZ), 67 Apollo Drive, Rosedale, North Shore 0632, New Zealand
(a division of Pearson New Zealand Ltd)
Penguin Books (South Africa) (Pty) Ltd, 24 Sturdee Avenue,
Rosebank, Johannesburg 2196, South Africa

Penguin Books Ltd, Registered Offices: 80 Strand, London WC2R 0RL, England

First published in Great Britain by Penguin Books Ltd 1996
First published in the United States of America by Viking Penguin,
a division of Penguin Books USA Inc. 1997
Published in Penguin Books 1998

20

Illustrations by Nancy Tolford

THE LIBRARY OF CONGRESS HAS CATALOGUED THE VIKING AMERICAN EDITION AS FOLLOWS:
Ridley, Matt.
The origins of virtue: human instincts and the evolution of cooperation / Matt Ridley.
p. cm.
Includes bibliographical references and index.
ISBN 0-670-87449-3 (hc.)
ISBN 978-0-14-026445-6 (pbk.)
1. Evolution (Biology) 2. Altruism. 3. Ethics, Evolutionary. I. Title.
QH366.2.R527 1997
303.5—dc20 96-44907

Printed in the United States of America
Set in Sabon

Contents

Acknowledgements

The words in this book are entirely my own; but the insights and ideas belong mostly to other people. My greatest debt is to those who shared their thoughts and discoveries with me so generously. Some submitted to long interrogations or sent papers and books, some gave moral or practical support and some read or criticized drafts of chapters. I thank them all sincerely.

They include: Terry Anderson, Christopher Badcock, Roger Bate, Laura Betzig, Roger Bingham, Monique Borgehoff Mulder, Mark Boyce, Robert Boyd, Sam Brittan, Stephen Budiansky, Stephanie Cabot, Elizabeth Cashdan, Napoleon Chagnon, Bruce Charlton, Dorothy Cheney, Jeremy Cherfas, Leda Cosmides, Helena Cronin, Lee Cronk, Clive Crook, Bruce Dakowski, Richard Dawkins, Robin Dunbar, Paul Ekman, Wolfgang Fikentscher, Robert Frank, Anthony Gottlieb, David Haig, Bill Hamilton, Peter Hammerstein, Garrett Hardin, John Hartung, Toshikazu Hasegawa, Kristen Hawkes, Kim Hill, Robert Hinde, Mariko Hiraiwa-Hasegawa, David Hirshleifer, Jack Hirshleifer, Anya Hurlbert, Magdalena Hurtado, Lamar Jones, Hillard Kaplan, Charles Keckler, Bob Kentridge, Desmond King-Hele, Mel Konner, Robert Layton, Brian Leith, Mark Lilla, Tom Lloyd, Bobbi Low, Michael McGuire, Roger Masters, John Maynard Smith, Gene Mesher, Geoffrey Miller, Graeme Mitchison, Martin Nowak, Elinor Ostrom, Wallace Raven, Peter Richerson, Adam Ridley, Alan Rogers, Paul Romer, Garry Runciman, Miranda Seymour, Stephen Shennan, Fred Smith, Vernon Smith, Lyle Steadman, James Steele, Michael Taylor, Lionel Tiger, John Tooby, Robert Trivers, Colin Tudge, Richard Webb, George Williams, Margo Wilson and

Robert Wright. It has been a privilege to see these minds at work and I only hope I have done justice to their ideas.

For their patience and advice, I thank my agents, Felicity Bryan and Peter Ginsberg; my editors and encouragers at Viking Penguin – Ravi Mirchandani, Clare Alexander and Mark Stafford; and several newspaper and magazine editors who gave me the space to try out some ideas in print – Charles Moore, Redmond O'Hanlon, Rosie Boycott and Max Wilkinson.

Above all, and for everything, I thank my wife, Anya Hurlbert.

Prologue

In which a Russian anarchist escapes from prison

> I was in pain to consider the miserable condition of the old man; and now my alms, giving some relief, doth also ease me.
>
> Thomas Hobbes, explaining why he gave sixpence to a beggar.

The prisoner was in a dilemma. As he paced slowly along his accustomed path, he suddenly heard a violin, in the open window of a house overlooking the prison yard. It was playing an exciting Kontski mazurka. The signal! But he was at that point in his walk farthest from the prison gate. His escape plan must work the first time or not at all, for it depended upon surprising the guards.

Now he had to shed his heavy dressing-gown, turn and run towards the open gate of the prison before the guards could catch him. The gate was open to receive a regular delivery of firewood. Once outside, his friends would whisk him away through the streets of St Petersburg in a carriage. The plans had been carefully laid, and relayed to the prisoner in cipher in a message hidden in a watch delivered to him by a woman visitor. His friends were posted along the street for two miles, each giving a different signal to the next that the streets were clear of traffic. The violin was the signal that the street was clear, the carriage was in place, the guard at the hospital gate close to the carriage was being engaged in deep, misleading conversation by the prisoner's confederate on the subject of how parasites appear under the microscope (research had revealed that the guard's hobby was microscopy), and that all was ready.

But one slip and he would never have another chance. He would probably be returned from the St Petersburg military hospital jail to the dark, damp, enfeebling gloom of the Peter and Paul fortress, where he had already spent two lonely, scurvy-ridden years. So he must choose his moment carefully. Would the mazurka continue until he reached the point in the path nearest the prison gate? When should he run?

With trembling tread he paced back along the path towards the prison gate. He reached the end of the path and turned to look at the sentry who was following him: the man had stopped five paces behind. The violin was still playing (and well, he thought).

Now! With the two quick motions he had practised a thousand times, he flung off his cumbrous garment and broke into a run. The sentry gave chase, flinging his rifle forwards to strike the prisoner down with the bayonet. But desperation lent the prisoner strength and he reached the entrance unscathed and a few paces ahead of his pursuer. Through the gate he hesitated for a second on seeing that the carriage was occupied by a man in a military cap. Sold to the enemy! he thought. But then he noticed the sandy whiskers of his friend, the tsaritsa's personal physician and a secret revolutionary, beneath the cap; he leapt aboard. The cab sped away into the city, pursuit being hampered by his friends who had hired all nearby cabs. They drove to a barber's shop, shaved off the prisoner's beard and by evening were ensconced in one of the most fashionable of St Petersburg's restaurants, where the secret police would never even think of looking.

Mutual aid

Much, much later, the prisoner would remember the fact that he owed his freedom to the courage of others: the woman who brought the watch, the woman who played the violin, the friend who drove the carriage and the physician who sat inside it, all the various confederates who kept the streets clear of traffic while he made good his escape. It was a team effort that sprang him from jail, and the memory was to ignite in his mind a whole theory of human evolution.

Prince Peter Kropotkin is remembered today, if at all, as an anarchist. But his escape from a tsarist prison in 1876 was the most dramatic and notable moment in a long, controversial and public life. From an early age the prince had been marked out for distinction. The son of a distinguished aristocratic general, when only eight years old he was noticed by Tsar Nicholas I at a ball, where he was a page dressed in Persian costume, and ordered to join the Corps of Pages, Russia's most select military academy. In the Corps he excelled, and was picked as the sergeant, a post that carried the job of personal page to the tsar himself (by now Alexander II). A glittering military or diplomatic career lay before him.

But Kropotkin, a brilliant mind infected with free-thinking by a French tutor, had other ideas. Joining a scandalously unfashionable Siberian regiment, he spent several years exploring the far eastern reaches of Siberia, pioneering several new routes through the mountains and river gorges of that land and developing his own precocious theories about the geology and history of the Asian continent. He returned to St Petersburg a geographer of note and, disgusted by the political prisons he had seen, a secret revolutionary. After a visit to Switzerland, where he fell under the spell of the anarchist Michael Bakunin, he joined an underground circle of anarchists in the Russian capital, and worked to foment the revolution. Sometimes he went straight from dining at the Winter Palace to meetings where he could agitate in disguise among the workers and peasants. Under the pseudonym Borodin, he published inflammatory pamphlets and developed great renown as a firebrand speaker.

When the police eventually caught up with Borodin, and he was revealed to be none other than the renowned Prince Kropotkin, the tsar and all his court were shocked and furious. They were even more angry when, two years later, he escaped from prison in so flamboyant a manner and travelled undetected into exile. He lived successively in England, Switzerland, France and eventually, when nowhere else would take him, in England again. There he turned gradually from agitation to more judicious philosophical writing and speaking on behalf of the anarchist cause, and inveighing against the rival creed of Marxism, which he felt was intent on reinventing the centralized, autocratic, bureaucratic state he and others had fought so hard to undermine.

In 1888, balding, bearded, bespectacled, rotund and kindly, Kropotkin was living the life of an impoverished freelance writer in Harrow, on the outskirts of London, still patiently expecting the revolution in his native land. That year, stung by an essay of Thomas Henry Huxley's with which he disagreed, the anarchist began work on what was to prove his enduring legacy, the chief thing for which he is now remembered. It became a book, called *Mutual Aid: A Factor in Evolution*, and it is a prophetic work, though now largely forgotten.

Huxley argued that nature was an arena for pitiless struggle between self-interested creatures. This placed him in a long tradition, going back through Malthus, Hobbes, Machiavelli and St Augustine to the Sophist philosophers of Greece, which viewed human nature as essentially selfish and individualistic unless tamed by culture. Kropotkin appealed to a different tradition, derived from Godwin, Rousseau, Pelagius and Plato, that man was born virtuous and benevolent, but was corrupted by society.

Kropotkin argued that the emphasis Huxley placed upon the 'struggle for existence' simply did not accord with what he observed in the natural world, let alone in the world of men. Life was not a bloody free-for-all, or (in Huxley's paraphrase of Thomas Hobbes) 'a war of each against all', but was characterized as much by cooperation as by competition. The most successful animals, indeed, seemed to be the most cooperative. If evolution worked by pitting individuals against each other, it also worked by designing them to seek mutual benefit.[1]

Kropotkin refused to accept that selfishness was an animal legacy and morality a civilized one. He saw cooperation as an ancient, animal tradition with which man, like other animals, was endowed. 'But if we resort to an indirect test, and ask Nature "Who are the fittest: those [species] who are continually at war with each other, or those who support one another?" we at once see that those animals which acquire habits of mutual aid are undoubtedly the fittest.' He could not stomach the idea that life was a ruthless struggle of selfish beings. Had he not been sprung from prison by a dozen faithful friends at great risk to their own lives? Where in Huxley's struggle could he explain such altruism? Parrots are superior to other birds, he suggested, because they are more sociable and therefore more intelligent. And among people, cooperation is just as pronounced among primitive tribes as it is among civilized citizens. From a common meadow in a rural village to the structure of a medieval guild, Kropotkin argued, the more people helped each other, the more the community thrived.

> The sight of a Russian commune mowing a meadow – the men rivalling each other in their advance with the scythe, while the women turn the grass over and throw it up into heaps – is one of the most awe-inspiring sights; it shows what human work might be and ought to be.

Kropotkin's was not a mechanistic theory of evolution, like Darwin's. He could not explain how mutual aid gained such a foothold, except by the selective survival of sociable species and groups in competition with less sociable ones – which was just to remove competition and natural selection one step, to the group rather than the individual. But he had posed a question that reverberates through economics, politics and biology a century later. If life is a competitive struggle, why is there so much cooperation about? And why, in particular, are people such eager cooperators? Is humankind instinctively an anti-social or a pro-social animal? That is my quest in this book: the roots of human society. I shall demonstrate that Kropotkin was half right and those roots lie much deeper than we think. Society works not because we have consciously invented it, but because it is an ancient product of our evolved predispositions. It is literally in our nature.[2]

Original virtue

This is a book about human nature, and in particular the surprisingly social nature of the human animal. We live in towns, work in teams, and our lives are spiders' webs of connections – linking us to relatives, colleagues, companions, friends, superiors, inferiors. We are, misanthropes notwithstanding, unable to live without each other. Even on a practical level, it is probably a million years since any human being was entirely and convincingly self-sufficient: able to survive without trading his skills for those of his fellow humans. We are far more dependent on other members of our species than any other ape or monkey. We are more like ants or termites who live as slaves to their societies. We define virtue almost exclusively as pro-social behaviour, and vice as anti-social behaviour. Kropotkin was right to emphasize the huge role that mutual aid plays in our species, but wrong and anthropomorphic to assume that therefore it applied to other species as well. One of the things that marks humanity out from other species, and accounts for our ecological success, is our collection of hyper-social instincts.

Yet to most people instincts are animal things, not human. The conventional wisdom in the social sciences is that human nature is simply an imprint of an individual's background and experience. But our cultures are not random collections of arbitrary habits. They are canalized expressions of our instincts. That is why the same themes crop up in all cultures – themes such as family, ritual, bargain, love, hierarchy, friendship, jealousy, group loyalty and superstition. That is why, for all their superficial differences of language and custom, foreign cultures are still immediately comprehensible at the deeper level of motives, emotions and social habits. Instincts, in a species like the human one, are not immutable genetic programmes; they are predispositions to learn. And to believe that human beings have instincts is no more determinist than to believe they are the products of their upbringings.

It is the claim of this book that the answer to an old question – how is society possible? – is suddenly at hand, thanks to the insights of evolutionary biology. Society was not invented by reasoning men. It evolved as part of our nature. It is as much a product of our genes as our bodies are. To understand it we must look inside our brains at the instincts for creating and exploiting social bonds that are there. We must also look at other animals to see how the essentially competitive business of evolution can sometimes give rise to cooperative instincts. This book is on three levels. It is about the billion-year coagulation of our genes into cooperative teams, the million-year coagulation of our ancestors into cooperative societies, and the thousand-year coagulation of ideas about society and its origins.

This is an impossibly immodest task, and I make no claim on having the last word on any of these matters. I cannot even be confident that many of the ideas I discuss in this book are right. But I shall be satisfied if some of them prove to have led in the right direction. My aim is to convince you to try to step out of your human skin and look back at our species with all its foibles. Naturalists know that each species of mammal can be distinguished as easily from another by its behaviour as by its appearance, and I am convinced that human beings are the same. We have idiosyncratic, species-specific ways of behaving that distinguish us from chimpanzees and bottlenose dolphins – we have, in short, an evolved nature. It sounds obvious when I put it like that but we so rarely do put it like that. We are always comparing ourselves with ourselves, a dismally narrow perspective. Suppose, therefore, you have been commissioned to write a book on life on earth, perhaps for a Martian publisher. You are devoting a chapter to each species of mammal (it will be a long book), giving a description of not just its body shape, but its behaviour as well. You have reached the apes and now have before you the job of describing *Homo sapiens*. How would you characterize the behaviour of this funny-looking large ape? One of the first ideas that would come to mind is 'social: lives in large groups with complex inter-relationships among individuals'. It is that which is the theme of my book.

CHAPTER ONE

The Society of Genes

In which there is a mutiny

The society formed by the hive bee fulfils the ideal of the communistic aphorism 'to each according to his needs, from each according to his capacity'. Within it, the struggle for existence is strictly limited. Queen, drones and workers have each their allotted sufficiency of food . . . A thoughtful drone (workers and queens would have no leisure for speculation) with a turn for ethical philosophy, must needs profess himself an intuitive moralist of the purest water. He would point out, with perfect justice, that the devotion of the workers to a life of ceaseless toil for a mere subsistence wage, cannot be accounted for either by enlightened selfishness, or by any other sort of utilitarian motives.

T. H. Huxley, *Evolution and Ethics. Prolegomena*, 1894

'The ants and termites,' wrote Prince Kropotkin, 'have renounced the "Hobbesian war", and they are the better for it.' If ever there was proof of the power of cooperation, ants, bees and termites are it. There are probably ten thousand billion ants on the planet, weighing in aggregate as much as all the human beings put together. It has been estimated that three-quarters of all the insect biomass – and in some places one-third of all the animal biomass – in the Amazon rain forest consists of ants, termites, bees and wasps. Forget the vaunted biodiversity of the millions of beetle species. Forget monkeys, toucans, snakes and snails. The Amazon is dominated by colonies of ants and termites. You can detect the formic acid that ants emit from an aeroplane overhead. They are perhaps even more ubiquitous in deserts. Were it not for an inexplicable intolerance for cool temperatures, ants and termites would prevail in temperate climates as well. As much as ourselves, they are the masters of the planet.[1]

The beehive and the ant's nest have been the favourite metaphor of human collaboration since time immemorial. To Shakespeare, a hive was a benevolent despotism, living in harmonious obedience to a monarch. As the Archbishop puts it, sycophantically, to Henry V:

> For so work the honey-bees,
> Creatures that by a rule in nature teach
> The act of order to a peopled kingdom.
> They have a king, and officers of sorts,
> Where some, like magistrates, correct at home;

Others, like merchants, venture trade abroad;
Others, like soldiers, armèd in their stings,
Make boot upon the summer's velvet buds,
Which pillage they with merry march bring home
To the tent-royal of their emperor;
Who, busied in his majesty, surveys
The singing masons building roofs of gold,
The civil citizens kneading up the honey,
The poor mechanic porters crowding in
Their heavy burdens at his narrow gate,
The sad-eyed justice, with his surly hum,
Delivering o'er to executors pale
The lazy yawning drone.

In short, the beehive was hierarchical Elizabethan society, writ small.

Four centuries later some anonymous polemicist saw it differently. As Stephen Jay Gould relates:

One day, at the New York World's Fair in 1964, I entered the Hall of Free Enterprise to escape the rain. Inside, prominently displayed, was an ant colony bearing the sign: 'Twenty million years of evolutionary stagnation. Why? Because the ant colony is a socialist, totalitarian system.'[2]

What these two descriptions have in common is not just an instinctive comparison between the societies of social insects and human beings but a recognition that somehow the ants and the bees are better than us at doing something we strive towards. Their societies are more harmonious, more directed towards the common, or greater, good, whether it be communism or monarchy.

A single ant or honey bee is as feeble and doomed as a severed finger. Attached to its colony, though, it is as useful as a thumb. It serves the greater good of its colony, sacrificing its reproduction and risking its life on behalf of its colony. Ant colonies are born, grow large, reproduce and die, just like bodies. In the harvester ant of Arizona, the queen lives for fifteen or twenty years. In the first five years of her life, the colony grows until it reaches about 10,000 workers. Between the ages of three and five the colony goes through a period of what one researcher calls 'obnoxious adolescence', when it attacks and challenges neighbouring colonies, just like an adolescent ape establishing itself in the troop hierarchy. At five, the colony ceases to grow, like a mature ape, and begins to produce winged reproductives instead: the equivalents of a body's sperm and eggs.[3]

As a result of their collective holism, ants, termites and bees can indulge in ecological strategies that would be impossible for solitary creatures. Bees can search out the nectar of evanescent flowers, directing each other to the best feeding grounds; ants can likewise scavenge with frightening efficiency, calling voluminous recruits to an open jar of jam in a few short minutes. The beehive is like a single many-tentacled creature, dipping its fingers into flowers a mile or more from its nest. Some termites and ants build towering nests and deep, underground chambers in which to raise agricultural crops of fungus on carefully prepared compost of chopped leaves. Others farm aphids like racketeering dairymen, extracting the honeydew in exchange for protection. Others, more viciously, raid each other's nests to raise armies of slave workers duped into caring for the wrong species. Some carry on collective warfare against rival colonies. The safari ants of Africa swarm across the countryside in armies 20 million strong and 20 kilograms in aggregate weight, spreading terror as they go and devouring every living thing not fast enough to escape, even small mammals and reptiles. The ant, the bee and the termite represent the triumph of collective enterprise.

If ants dominate tropical forests on land, even more collective animals are even more dominant in the most diverse marine ecosystems: corals. In the submarine equivalent of the Amazon rain forest, the Great Barrier Reef of Australia, colonial animals form not only the dominant animals but the very trees as well – the primary producers. Corals build the reef, fix the carbon using their sunlight-powered confederate algae and consume the animals and plants of the water column, their stinging tentacles always sieving the water for algae and small invertebrates. Corals are collectives, like ant colonies, the only difference being that the individual animals that make up the collective are fixed in a permanent embrace, rather than free to come and go as individuals. The individuals may die, but the colony is close to immortal. Some coral reefs have been continuously alive for more than 20,000 years and lived through the last ice age.[4]

The first life on earth was atomistic and individual. Increasingly, since then, it has coagulated. It has become a team game, not a contest of loners. By 3.5 billion years ago there were bacteria five-millionths of a metre long and run by a thousand genes. Even then there was probably teamwork. Today some bacteria swarm together to build 'fruiting bodies' to disperse their spores. Some blue-green algae – simple bacteria-like life forms – form colonies, with even the rudiments of a division of labour between cells. By 1.6 billion years ago there were complex cells a million times heavier than bacteria and run by teams of 10,000 genes or more: the protozoa. By 500 million years ago there were complex bodies of animals comprising a billion cells; the largest animal on the planet was a trilobite – an arthropod the size of a mouse. Ever since then the biggest bodies have been getting bigger and bigger. The largest plants and animals that have ever lived on earth – the giant sequoia and the blue whale – are alive today. A blue whale has 100,000 trillion cells in its body. But already a new form of coagulation is occurring: social coagulation. By 100 million years ago there were complex colonies of ants run by teams of a million bodies or more and now they are among the most successful designs on the planet.[5]

Even the mammals and birds are beginning to coagulate socially. Florida scrub jays, splendid fairy wrens and green woodhoopoes, among other birds, breed cooperatively: a male, a female and several full-grown young share the duties of caring for the newest offspring. Wolves, wild dogs and dwarf mongooses do much the same – delegating reproduction to the senior pair within the troop. In one especially bizarre case, a burrowing mammal has produced something approaching a termite nest. The naked mole rat of East Africa lives in underground colonies of seventy or eighty animals, one of which is a gigantic queen and the others diligent, celibate workers. Like termites or bees, mole-rat workers even risk their lives on behalf of their colonies, by, for instance, rushing to block a tunnel when a snake invades it.[6]

The inexorable coagulation of life continues. Ants and corals are inheriting the earth. Mole rats may one day be as successful. Where will it stop?[7]

The Russian doll of collaboration

Floating through the oceans as predatory as a swarm of safari ants, the Portuguese man-o'-war, *Physalia*, with its sixty-foot stinging tentacles, its wind-powered sail-float, in menacing baby blue, and its fearsome reputation, is not an animal but a commune. It consists of thousands of tiny individual animals stitched together and sharing a common fate. Like ants in a colony each animal knows its place and its duty. Gastrozooids are the workers, collecting food, dactylozooids are the soldiers, defending the colony, gonozooids are the queens, reproducing.

Through the halls of Victorian zoology, an argument raged about *Physalia*. Was it a colony or an animal? Thomas Henry Huxley, dissecting it aboard HMS *Rattlesnake*, maintained that it was nonsense to call the zooids individual animals. They were just organs of a body. We now consider he was wrong, because each zooid is derived from a complete, little multicellular organism. But, though he was wrong about the zooid's history, Huxley was right in a philosophical sort of way. The zooids cannot live alone. They are as much dependent on the colony as my arm is on my stomach. The same, argued William Morton Wheeler in 1911, applies to an ant colony. It is an organism, with soldiers instead of an immune system, queens instead of ovaries and workers instead of a stomach.

This debate missed the point. The point is not that a Portuguese man-o'-war or an ant colony is really a single organism; it is that each single organism is a collective. It consists of millions of individual cells, each in its own way self-sufficient, but also dependent on the whole, just like a worker ant. The question that we should be asking is not why do some bodies get together to form colonies, but why do cells get together to form bodies? A shark is just as much a collective as a man-o'-war, only it is a collective of a million billion collaborating cells, whereas the man-o'-war is a collective of collectives of cells.

The organism itself needs explaining. Why do its cells gang together? The first person to see this really clearly was Richard Dawkins in his book *The Extended Phenotype*. If cells were lit up as little lights, he pointed out, we would see, when a person walked past, 'a million billion glowing pinpricks move in unison with each other and out of step with all the members of other such galaxies.'[8]

There is nothing in principle that stops cells working alone: many do, successfully, as amoebae and other protozoa. In one especially strange case, the creature can be either a single cell or a fungus-like growth. The slime mould consists of a group of about 100,000 amoebae that go their separate ways until conditions become unpromising. Then the cells all gather together in a mound, the mound grows taller, falls over and then sets off as a 'slug' the size of a grain of rice, looking for pastures new. If it fails, the slug adopts the shape of a Mexican hat, from the centre of which a ball of cells gradually grows upwards, supported by a long and slender stalk. The ball hardens into 80,000 spores, which wave in the wind, hoping to catch the body of a passing insect that can unwittingly transport them to a better place to start new colonies of independent amoebae elsewhere. The 20,000 stalk cells just die, martyrs to the fraternal welfare of the spores.[9]

These slime moulds are confederations of separate cells, quite capable both of living alone and of gathering together to make a temporary organism. But look closer and notice that even cells are collectives. They are formed from the symbiotic collaboration between bacteria, or so most biologists believe. Every cell in your body is home to mitochondria, tiny bacteria so specialized as energy-producing batteries that about seven or eight hundred million years ago they surrendered their independence in exchange for a comfortable life inside the cells of your ancestors. Even your cells are coalitions.

Nor need we leave the Russian doll there. For inside the mitochondria are little chromosomes, carrying genes, and inside the nuclei of your cells are forty-six larger chromosomes, carrying more genes, perhaps 75,000 of them in all. Chromosomes go about in teams of twenty-three pairs in human beings, rather than alone. But they could be individual – as they are in bacteria. And chromosomes also are collaborations, not individuals: collaborations of genes. Genes can go about in tiny teams of fifty or so (in which case we call them viruses), but many choose not to. They team up to form whole chromosomes: teams of thousands of closely linked genes. Even genes may not be atomistic: some of them produce only partial messages which must be stitched together with messages from other genes to make sense.[10]

So the quest for collaboration has taken us unexpectedly deep into biology. Genes team up to form chromosomes; chromosomes team up to form genomes; genomes team up to form cells; cells team up to form complex cells; complex cells team up to form bodies; bodies team up to form colonies. A beehive is a collaborative enterprise on far more levels than first appears.

The selfish gene

There was a revolution in biology in the mid 1960s, pioneered especially by two men, George Williams and William Hamilton. This revolution is best known by Richard Dawkins's phrase 'the selfish gene', and at its core lies the idea that individuals do not consistently do things for the good of their group, or their families, or even themselves. They consistently do things that benefit their genes, because they are all inevitably descended from those that did the same. None of your ancestors died celibate.

Williams and Hamilton are both naturalists and loners. Williams, the American, began his career as a marine biologist; Hamilton, the Briton, as a student of the social insects. In the late 1950s and early 1960s first Williams and then Hamilton argued their way to a new and startling way of understanding evolution in general, and social behaviour in particular. Williams began by suggesting that to grow old and die was a rather counterproductive thing for a body to do, but that it made sense for the genes to programme obsolescence into the body after reproduction. Animals (and plants), he concluded, are designed to do things not for their species, or for themselves, but for their genes.

Usually the genetic and the individual interest coincide – but not always (salmon die with the effort of spawning; bees commit suicide in the act of stinging). Often the interest of the genes requires a creature to do things that benefit its offspring – but not always (birds desert their babies if food runs short; chimpanzee mothers heartlessly wean their imploring young from the teat). Sometimes it means doing things for the benefit of other relatives (ants and wolves help their sisters breed). Occasionally, it means doing things that benefit the larger group (musk oxen stand shoulder to shoulder against a wolf pack to protect the young). Sometimes it means making other creatures do things that are bad for them (colds make you cough; salmonella gives you diarrhoea). But always, without exception, living things are designed to do things that enhance the chances of their genes or copies of their genes surviving and replicating. Williams made the point with characteristic bluntness: 'As a general rule, a modern biologist seeing an animal doing something to benefit another assumes either that it is being manipulated by the other individual or that it is being subtly selfish.'[11]

This idea emerged from two directions. First, it came out of theory. Given that genes are the replicating currency of natural selection, it is an inevitable, algorithmic certainty that genes which cause behaviour that enhances the survival of such genes must thrive at the expense of genes that do not. It is just a simple consequence of the fact of replication. The insight also came out of observation and experiment. All sorts of behaviour that had seemed puzzling when seen through the lens of the individual or the species, suddenly became clear when seen through a gene-focused lens. In particular, as Hamilton triumphantly showed, the social insects, by helping their sisters to breed, left more copies of their genes in the next generation than by trying to breed themselves. From the gene's point of view, therefore, the astonishing altruism of the worker ant was purely, unambiguously selfish. The selfless cooperation of the ant colony was an illusion: each worker ant was striving for genetic eternity through its brothers and sisters, the queen's royal offspring, rather than through its own offspring, but it was doing so with just as much gene-selfishness as any human being elbowing aside his rivals on the way up the corporate ladder. The ants and termites might, as Kropotkin had said, have 'renounced the Hobbesian war' as individuals, but their genes had not.[12]

The mental impact of this revolution in biology for those close to it was dramatic. Like Copernicus and Darwin, Williams and Hamilton dealt a humiliating blow to human self-importance. Not only was the human being just another animal, but it was also the disposable plaything and tool of a committee of self-interested genes. Hamilton himself recalls the moment when it dawned upon him that his body and his genome were more like a society than a machine. 'There had come the realization that the genome wasn't the monolithic data bank plus executive team devoted to one project – keeping oneself alive, having babies – that I had hitherto imagined it to be. Instead, it was beginning to seem more a company boardroom, a theatre for a power struggle of egotists and factions . . . I was an ambassador ordered abroad by some fragile coalition, a bearer of conflicting orders from the uneasy masters of a divided empire.'[13]

Richard Dawkins, coming upon the same ideas as a young scientist, was equally stunned: 'We are survival machines – robot vehicles blindly programmed to preserve the selfish molecules known as genes. This is a truth which still fills me with astonishment. Though I have known it for years, I never seem to get fully used to it.'[14]

Indeed, for one of Hamilton's readers the impact of the idea of the selfish gene was tragic. George Price taught himself genetics in order to disprove Hamilton's stark conclusion that altruism was just genetic selfishness, but instead proved it indisputably correct – indeed, even improved the algebra and made some important contributions to the theory himself. The two began to collaborate, but Price, who was showing increasing signs of mental instability, turned to religion for solace, gave away all his possessions to the poor and committed suicide in a bare and cold London squat, some letters from Hamilton among his few possessions.[15]

A much more common reaction has been to hope that Williams

and Hamilton will go away. The very phrase 'selfish gene' sounded sufficiently Hobbesian to repel most social scientists from the selfish-gene revolution, and drive more conventional evolutionary biologists like Stephen Jay Gould and Richard Lewontin into a perpetual rear-guard fight against it. Like Kropotkin, they were repelled by the notion (actually a misunderstanding, as we shall see) that Williams, Hamilton and their colleagues were trying to reduce all selflessness to fundamental self-interest. They thought that was, to paraphrase Friedrich Engels, to drown the richness of nature in the icy waters of self-interest.[16]

The selfish embryo

Yet the selfish-gene revolution, far from being a bleak and Hobbesian injunction to go out and ignore the good of others, is in fact the very opposite. It makes room for altruism after all. For, whereas Darwin and Huxley, like classical economists, had perforce assumed that people act out of self-interest, Williams and Hamilton have come to the rescue by revealing a much more powerful engine of behaviour: genetic interest. Selfish genes sometimes use selfless individuals to achieve their ends. Suddenly, therefore, altruism by individuals can be understood. Huxley, by thinking only in terms of individuals, was fixated on the struggle between them and missed, as Kropotkin pointed out, the myriad ways in which individuals do not always fight each other. Had he known to think in terms of genes, he might have reached a less Hobbesian conclusion about individuals. As we shall see later, biology softens economic lessons rather than hardens them.

The genetic perspective echoes an old argument about motives. If a mother is selfless towards her offspring only because her genes are being selfish, she is still, as an individual, behaving selflessly. If we know that an ant is altruistic only because its genes are egotistical, we still cannot deny that the ant itself is altruistic. If we can allow that individual people are nice to each other, then the 'motives' of the genes that cause the virtue can go hang. Pragmatically, it does not matter to us that a man saves a drowning companion because he wants the glory rather than because he wants to do good. Likewise, it does not matter that he is under the orders of his genes, rather than choosing a course of action of his own free will. The deed is what counts.

Some philosophers have argued that there cannot be such a thing as animal altruism, because altruism must imply a generous motive rather than a generous act. Even St Augustine wrestled with this question; alms giving, he said, must be done for the motive of love of God, not out of pride. A similar question divided Adam Smith from his teacher, Francis Hutcheson, who argued that benevolence motivated by vanity or self-interest was not benevolence. Smith thought this too extreme. A man may do a good deed, even if he does it out of vanity. More recently, the economist Amartya Sen, echoing Kant, has written:

> If the knowledge of torture of others makes you sick, it is a case of sympathy ... It can be argued that behaviour based on sympathy is in an important sense egoistic, for one is oneself pleased at others' pleasure and pained at others' pain, and the pursuit of one's own utility may thus be helped by sympathetic action.[7]

In other words, the more you truly feel for people in distress, the more selfish you are being in alleviating that distress. Only those who do good out of cold, unmoved conviction are 'true' altruists.

Yet what matters to society is whether people are likely to be nice to each other, not their motives. If I am setting out to raise money for a charity, I am not going to return the cheques of companies and celebrities on the grounds that they are motivated more by the search for good publicity than by the cause itself. Likewise, when Hamilton developed the theory of kin selection, he did not for one moment interpret the worker ant as selfish, rather than selfless, because it remained sterile. He merely interpreted its selfless behaviour as a consequence of selfish genes.

Consider, for example, inheritance. All over the world one of the incentives people have to earn wealth is to leave it to their children. There is no extinguishing this human instinct: with relatively few exceptions, people try to pass on much of their wealth to the next generation rather than spend it all, give it to charity or just relinquish it to be shared with strangers on their death. Yet there is no place in classical economics for such a generous motive, obvious though it is. Economists have to accept it and assume it, but they cannot explain it because it brings no benefit to the individual. In a gene-centred view of humankind, however, such astonishing altruism makes perfect sense, for the money is following the genes, even if it abandons the individuals.

If the selfish gene saves Rousseau from the clutches of the Hobbesians, it is by no means entirely friendly to the angels. For it also predicts that universal benevolence is impossibly Utopian, that the fungus of selfishness will be ready to strike at the heartwood of any harmonious whole. It will lead us to suspect self-interest to be the cause of endless mutinies. Just as Hobbes argued that the state of nature was not one of harmony, so Hamilton and Robert Trivers, two pioneers of selfish-gene logic, argued that the relationships between parents and offspring, or between mates, or between social partners was not one of mutual satisfaction, but one of mutual struggle to exploit the relationship.

Take the case of a foetus in the womb. Nothing could be more common than the interest between a mother and her foetus. She wants to bear it to term because it carries her genes into the next generation. It wants her to thrive, because otherwise it will die. They both use her lungs to get oxygen, they both depend on her heart to keep beating. The relationship is entirely harmonious; pregnancy is a cooperative effort.

Or so biologists used to think. Then, after Robert Trivers noticed how much conflict there routinely is between mother and infant after birth (over such matters as the timing of weaning), David Haig extended this thinking back into the womb. Consider, he said, the ways in which mother and foetus are not at one with each other. The mother wants to live to have another child; the foetus would prefer that she devote most of her effort to itself. The mother shares only half of the genes in the foetus and vice versa; if one of them has to die that the other may live, each would rather be the survivor.[18]

At the end of 1993 Haig published startling evidence against the conventional rosy view. In all sorts of ways, he found, the foetus and its slave, the placenta, act more like subtle internal parasites than like friends, trying to assert their interests over those of the mother. Cells from the foetus invade the artery supplying maternal blood to the placenta, embed themselves in the walls and destroy the muscle cells there, thus removing the mother's control over constriction of that artery. The high blood pressure and pre-eclampsia that often complicate pregnancy are caused largely by the foetus, using hormones to try to divert the mother's blood to itself by reducing the flow through her other tissues.

Likewise, there is a battle over blood sugar. During the last three months of pregnancy, a mother generally has stable levels of blood sugar, yet she is producing more and more insulin every day – insulin being a hormone that normally suppresses blood sugar levels. The reason for this paradox is simple: the placenta, under foetal control, secretes into its mother's blood increasing quantities of a hormone called human placental lactogen (hPL), which blocks the effect of insulin. Comparatively vast quantities of this hormone are produced during a normal pregnancy, although in occasional cases, where none is produced, neither mother nor foetus is any the worse. So both the foetus and the mother are churning out escalating quantities of hormones which have opposite effects and simply cancel each other out. What's going on?

In Haig's view it is a tug of war between a greedy foetus, trying to increase the amount of sugar in the mother's blood to feed itself, and a thrifty mother, trying to ensure that the foetus does not take too much of her precious blood sugar. In some women, the effect of this brief and stalemated war is to cause gestational diabetes – the foetus having won the battle too well. Moreover, the hPL hormone the foetus makes is dictated by a gene that it inherits from the father alone, as if the foetus were a paternal parasite inside the mother. What price harmony in the womb now?

Haig's point is not to try to claim that all pregnancies are tugs of bitter war between enemies; mother and child are still basically cooperating in the business of rearing the child. The mother is still astonishingly selfless as an individual in the way she nurtures and protects her children. But, as well as the shared genetic interest between them, there are also some divergent genetic ambitions. The mother's selflessness conceals the fact that her genes act as if motivated entirely by selfishness, whether being nice to the foetus or fighting it. Even within the inner sanctum of love and mutual aid – the womb itself – we have found ruthless assertion of self-interest.[19]

Mutiny in the beehive

The same pattern of conflict in the midst of cooperation is to be found in every other natural collaboration. At every stage there is the threat of mutiny, of rebellious individualism that might destroy the collective spirit.

Consider the question of celibate worker bees. Unlike many ants, worker bees are not incapable of producing young, yet they almost never do. Why not? Why does a worker not rebel against the tyranny of rearing her mother's other daughters, and have babies herself? It is not an idle question. In one hive in Queensland exactly that happened recently. A few of the workers began to lay eggs in a compartment separated from the rest of the hive by a queen excluder (a sieve the large-bodied queen cannot pass through). The eggs hatched into males (drones), which is not surprising since the workers had not mated, and eggs that have not been fertilized by a male are, in ants, bees and wasps, automatically male – such is the simple mechanism of sex determination in these insects.

If you ask a worker honey bee, 'Who would you prefer to be the mother of the hive's males?' her answer would be herself, her queen and only then another (randomly picked) worker – in that order: for that is the order of decreasing relatedness. The reason is that a honey bee queen mates with fourteen to twenty males and mixes their sperm thoroughly. Therefore, most workers are half-sisters of each other, not full sisters. A worker shares half her genes with her own son, a quarter of her genes with the queen's sons and less than a quarter with the sons of most other workers who are her half-sisters.

Therefore, each worker that lays its own eggs makes a greater contribution to posterity than a worker that desists. It follows that, in a few generations, breeding workers will inherit the world. What stops it happening?

Each worker prefers its own sons to the queen's; but equally each worker prefers the queen's sons to the sons of any other worker. So workers police the system themselves thereby incidentally serving the greater good. They are careful not to let each other breed in 'queen-right' colonies; they simply kill the offspring of other workers. Any egg not marked with a special pheromone by the queen is eaten by the workers. In the exceptional Australian hive, scientists concluded that one drone had passed on to some of the workers in the hive a genetic ability to evade the policing mechanism and to lay eggs that would not be eaten. A sort of majoritarianism, a parliament of the bees, normally keeps the workers from breeding.

Queen ants solve the problem in a different way: they produce workers who are physiologically sterile. Unable to reproduce, the workers cannot rebel, so there is nothing to require the queen to mate with many males. All workers are full sisters. They would prefer workers' sons to queens' sons, but they cannot make them. Another exception that also proves the rule is found among the bumble bees, or humble bees. 'Kill me a red-hipped humble bee on the top of a thistle;' said Bottom to Cobweb in *A Midsummer Night's Dream*, 'and, good monsieur, bring me the honeybag.' Following Bottom's example is not a commercial proposition. Humble bees, or bumble bees as they are nowadays known, do not produce honey in sufficient quantities to satisfy beekeepers. Elizabethan boys knew that they could raid a bumble bees' nest for the little waxen thimble of honey put aside for the queen's use on rainy days, but nobody ever kept a hive of bumble bees. Why not? They are just as industrious as honey bees. The answer is simple enough. A bumble bee colony never gets very large. At the most it may have four hundred workers and drones, nothing like the thousands of honey bees in a hive. At the end of the season the queens disperse to hibernate alone, starting afresh next year; no workers go with them.

There is a reason for this difference between bumble bees and honey bees, a curious and newly discovered one. Bumble bee queens are monogamous; each mates with only one drone. Honey bee queens are polyandrous, mating with many drones. The result is an odd piece of genetic arithmetic. Remember that male bees of all species are grown from unfertilized eggs, so all males are pure clones of half their mothers' genes. Workers, in contrast, have a father and a mother and are all female. Bumble bee workers are more closely related to the offspring of their sister workers (37.5 per cent, to be exact) than to the sons of their mother (25 per cent). Therefore, when the colony begins to produce males, the workers conspire not – like honey bees – with the queen and against their sisters, but with their sisters and against the queen. They rear workers' sons instead of royal sons. It is this disharmony between the queen and the workers that explains the bumble bees' smaller colonies which break up at the end of each season.[20]

The collective harmony of the hive is achieved only by suppressing selfish mutiny of individuals. The same applies to the collective harmony of the body, the cell, the chromosome and the genes. In the slime mould, the confederation of amoebae that comes together to build a stalk from which to launch spores, there is a classic conflict of interest. Up to a third of the amoebae will have to make the stalk, as opposed to the spores, and will die. An amoeba that avoids being in the stalk therefore thrives at the expense of a more public spirited colleague, and leaves more of its selfish genes behind. How does the confederation persuade the amoebae to do their stalk duty and die? Often the amoebae that come together to make a stalk are from different clones, so nepotism is not the only answer. Selfish clones might still prevail.

The question turns out to be a familiar one for economists. The stalk is a public good, provided for out of taxation – like a road. The spores are the private profits that can be made from using the road. The clones are like different firms who are facing the decision of how much tax to pay for the road. The 'law of equalization of net incomes' says that, knowing how many clones are contributing to the stalk, each clone should reach the same conclusion about how much to allocate to the spores (the net income). The rest should be paid in stalk (tax). It is a game in which cheating is suppressed, though precisely how is not yet clear.[21]

In human beings, too, there is always conflict between the selfish individual and the greater good. Indeed, so pervasive is this tendency that a whole theory of political science has come to be based upon it. Public-choice theory, devised by James Buchanan and Gordon Tullock in the 1960s, holds that politicians and bureaucrats are not exempt from self-interest. Although they may be charged with pursuing public duty rather than their own advancements and rewards, they come inevitably and always to pursue what is best for themselves and their agency rather than for its clients or the taxpayers who fund it. They exploit induced altruism: they enforce cooperation and then defect. This may seen unduly cynical, but then the opposing view – that bureaucrats are selfless servants of the public good ('economic eunuchs', as Buchanan put it) – is unduly naive.[22]

As C. Northcote Parkinson put it, defining the famous 'Parkinson's Law' (which is an eloquent presaging of the same theory), 'An official wants to multiply subordinates, not rivals; and officials make work for each other.' With delicious irony, Parkinson described the quintupling of the number of civil servants in Britain's Colonial Office between 1935 and 1954, during which time the number and size of colonies to be administered shrank dramatically. 'It would be rational,' he wrote, 'prior to the discovery of Parkinson's Law, to suppose that these changes in the scope of the Empire would be reflected in the size of its central administration.'[23]

The rebellion of the liver

In ancient Rome there was a distinction between two classes of citizen, the Plebeians and the Patricians. With the expulsion of the Tarquins, Rome rejected monarchy and became a republic. But soon the Patricians began to monopolize political power, religious office and legal privilege. No Plebeian, however wealthy, was allowed to become a senator or a priest, nor could he sue a Patrician. Only joining the army to fight Rome's wars was open to him, a dubious privilege at best. In 494 BC, fed up with this injustice, the Plebeians effectively went on strike against further warfare. Promised by a hurriedly appointed dictator, Valerius, protection from their debts, they returned to work, quickly defeated the Aequi, Volsci and Sabines in succession, and came back to Rome. The ungrateful Senate promptly overturned Valerius' promise, whereupon the furious Plebeians encamped in military order on the Mons Sacer outside the city, a menacing presence. The Senate sent a wise man, Menenius Agrippa, to argue with them, and he told them a fable.

> Once upon a time the members of the body began to grumble because they had all the work to do, while the belly lay idle, enjoying the fruits of their labour; so the hands, mouth, and teeth agreed to starve the belly into submission, but the more they starved it the weaker they themselves became. So it was plain that the belly also had its work to do, which was to nourish the other members by digesting and redistributing the food received.

With this rather feeble apology for corrupt politicians, Menenius defused the rebellion. In exchange for the election of two tribunes from among the Plebeians, with the power to veto the punishment of a Plebeian, the army disbanded and order was restored.[24]

Your body is only a whole because of elaborate mechanisms to suppress mutiny. Think of it from the point of view of the liver in a woman's body. It works away for three-score years and ten detoxifying the blood and generally regulating the chemistry of the body for no reward. At the end it just dies and rots, forgotten. Meanwhile, right next door, a few inches away, the ovaries sit quietly and patiently, contributing nothing to the body except some rather unnecessary hormones, but scoop the jackpot of immortality by producing an egg that carries its genes on into the next generation. The ovaries are like parasites on the liver.

Using the arguments of nepotism derived from Hamilton's theory of kin selection, we can argue that the liver should not 'mind' the ovary's parasitism that much, because it is a clone of the ovary, genetically speaking. So long as the same genes survive through the ovary, it does not matter that those in the liver perish. That is the difference between the ovary and a liver parasite: the ovary shares the same genes as the liver. But imagine that one day a mutant cell appears in the liver that has a special property, a property of launching itself into the blood, travelling to the ovary and replacing the eggs therein with little copies of itself. Such a mutant would thrive at the expense of normal livers and would gradually spread. Within a few generations we would all be descended from our mothers' livers, not her (original) ovaries. The mutant liver cell is not deterred by the logic of nepotism, because when it first appears its genes are not shared by the ovaries.

This is an example from fantasy, not medicine, but it is closer to the truth than you might think. It is a rough description of cancer. Cancer is the failure of cells to stop replicating. Cells that continue to replicate indefinitely thrive at the expense of normal cells. So cancerous tumours, especially those that remain sufficiently generalized in appearance to metastasize – i.e. spread throughout the body – are bound to take over the body. To prevent cancer the body must therefore persuade every one of its million billion cells to obey the order to cease replicating when growth or repair is finished. This is not as easy as it sounds, because in the trillions of ancestral generations that came before, the one thing those cells never did was cease to divide – if they had they would not have been ancestral. Your liver cells come not from your mother's liver but from the egg in her ovary. The order to stop replicating and become a good liver cell is one they have never heard before in all of the two billion years of their immortal existence (during a woman's lifetime, her egg cells do not cease replicating, so much as pause in mid-replication until fertilized). Yet they must obey it first time or the body will succumb to cancer.

There are, fortunately, a great array of devices in place to ensure that the cells do obey, a massive chain of safety-catches and fail-safes that must malfunction if cancer is to break out. Only towards the end of life, or under assault from extreme radiation or chemical damage, do these mechanisms begin to fail (semi-deliberately: cancer begins to strike at a different age for every species). It is, however, no accident that some of the most dangerous cancers are transmitted by viruses. The rebellious cells of the tumour have found a way to spread, not by taking over the ovary but by jumping free in a virus capsule.[25]

The worm in the gall

Nor is this logic confined to cancer. Many of the disorders of old age can be profitably seen in this light. As your life plays out, there is inevitably selective survival of those cell lines that are good at surviving, which unavoidably includes those cell lines that are good at surviving at the expense of the body as a whole. It is not some evil design; it is an inevitability. Bruce Charlton, coining the term endogenous parasitism for this process, has argued that 'the organism can be conceptualised as an entity which will progressively self-destruct from the moment of its formation.' Ageing does not need explaining; staying so young does.[26]

In a developing embryo, the conflict between selfish cells and the greater good is an even greater danger. As the embryo grows any genetic mutant that makes its own cells take over the germ cells – the cells that will reproduce – is bound to spread at the expense of any other mutant. So development must be a scramble between selfish tissues for the prize of becoming the gonads. Why is it not?

According to one interpretation, the answer lies in two strange features of the life of an embryo: maternal predestination and germ-line sequestration. For the first few days of its life, the fertilized egg is shut down genetically. Its genes are not allowed to be transcribed; this radio silence is dictated by the mother's genes, which impose a sort of pattern on the embryo through the distribution of the products of her own genes. By the time the embryo's own genes are released from house arrest, their fate is largely determined. A short time later – in the human case, a mere fifty-six days after fertilization – the germ line is complete and isolated: the cells that will become the eggs or sperm of the adult are already segregated from the rest of the embryo. There they will remain unaffected by all the mutations, injuries and brainwaves that occur to all the other genes in the body. Nothing that happens to you after the fifty-sixth day of your prenatal life can directly affect the genes of your descendants, unless it affects your testicles or ovaries. Every other tissue is deprived of the opportunity to become an ancestor, and to deprive a tissue of the chance of becoming an ancestor is to deprive it of the opportunity to evolve at the expense of its rivals. The ambitions of the cells of the body are therefore bent to the will of the greater good. The mutiny is largely defeated. As one biologist put it, 'The impressive harmony of development reflects not the common interest of independent, cooperating agents but the enforced harmony of a well-designed machine.'[27]

Maternal predestination and germ-line sequestration make sense only as attempts to suppress a selfish mutiny of the cells. They occur only in animals, not in plants or fungi. Plants suppress the mutiny in other ways, by retaining the ability of any cell to become a reproductive one but using their rigid cell walls to prevent any cells moving throughout the body. Systemic cancer is not possible in plants. Fungi have a different approach: they have no cells at all and genes must play a lottery for reproductive rights.[28]

Selfish subversion threatens the coagulators inside the next Russian doll, too. Just as the body is an uneasy triumph of harmony over cellular egoism, so the cell itself is a delicate compromise of the same kind. Within each cell of your body are forty-six chromosomes, twenty-three from each of your parents. This is your 'genome', your team of chromosomes. They all work together in perfect harmony, dictating the work of the cell.

If, however, you are one of the two to three per cent of people unknowingly infected by a curious form of parasite, you might have a more jaundiced view of chromosomes. These parasites are called B chromosomes. In appearance they are identical to ordinary chromosomes, if perhaps a little smaller than average. But they do not travel in pairs, they contribute almost nothing to the functioning of the cell and they generally refuse to swap genes with other chromosomes. They are simply along for the ride. Because they require the usual complement of chemical resources, they generally slow down the rate of growth of the creatures they inhabit or reduce their fertility and health. They have been little studied in human beings, but in at least one case they are known to delay fertility in women. In many other animals and plants they are more numerous and more obvious in their deleterious effects.[29]

Why, then, are they there at all? Biologists have exercised their ingenuity to answer this question. Some argue that they are there to promote variability among genes. Others argue that they are there to suppress variability among genes. Neither argument is convincing. The truth is, B chromosomes are parasites. They thrive not because they are good for the cells they inhabit but because they are good for themselves. They are particularly cunning at accumulating in reproductive cells, and even then they leave nothing to chance. When the cell divides to form an egg it randomly discards half of the genes (which will be replaced by the genes from the fertilizing sperm), depositing them in so-called polar bodies. B chromosomes, craftily and mysteriously, almost never get put in the polar bodies. So although animals and plants with B chromosomes are less likely to survive and breed than those without, B chromosomes are more likely to appear in their offspring than other genes. B chromosomes are chromosomal mutineers: egoists subverting the harmony of the genome.[30]

Within each chromosome, too, there is mutiny. Inside your mother's ovaries an elegant card game known as 'meiosis' took place to form the egg that was half of you. The dealer first shuffled and then cut the pack of cards that is her genes. Half the pack was then discarded, leaving the other half to become half of you. Each gene took its chance in the game, a fifty–fifty gamble on getting into the egg. With magnificent grace, the losers accepted their extinction and wished their more fortunate fellows well on their journey towards eternity.

However, were you a mouse or fruit fly you might have inherited a gene called a segregation distorter, which simply cheated in the card game. It has a way of ensuring that it gets into the egg or sperm, however the cards are cut. Segregation distorters, like B chromosomes, serve no useful function for the greater good of the mouse or fly. They serve only themselves. Because they are so good at spreading, they thrive even if they do harm to their host bodies.

They are mutineers against the prevailing order. They reveal the tension beneath the apparent harmony of the genes.

The greater good

Yet these phenomena are rare. What stops the mutiny? Why do segregation distorters, B chromosomes and cancer cells not succeed in winning the contest? Why does harmony generally prevail over selfishness? Because the organism, the coagulation, asserts its greater interest. But what is the organism? There is no such thing. It is merely the sum of its selfish parts; and a group of units selected to be selfish cannot surely turn altruistic.

The resolution of this paradox takes us back to the honey bees. Each worker bee has a selfish interest in producing drones; but each worker equally has a selfish interest that no other worker produce drones. For every selfish drone-producer there are thousands of bees with a selfish interest in preventing that drone production. So a bee hive is not, as Shakespeare thought, a despotism, run from above. It is a democracy, in which the individual wishes of the many prevail over the egoism of each.

Exactly the same applies to cancer cells, outlaw embryo tissues, segregation distorters and B chromosomes. Mutations that make genes suppress the selfishness of other genes are just as likely to thrive as selfish mutants. And there are far more places such mutations can occur: for every selfish mutation at one place, there are tens of thousands of other genes which will thrive if they accidentally stumble on mechanisms that cause the suppression of the selfish mutant. As Egbert Leigh has put it, 'It is as if we had to do with a parliament of genes: each acts in its own self-interest, but if its acts hurt others, they will combine together to suppress it.'[31] In the case of segregation distorters, the selfishness is averted by the division of the genome into many chromosomes and by 'crossing over' within each chromosome, which swaps genes back and forth and thus has the effect of separating a segregation distorter from the safety mechanism that prevents it destroying itself. These measures are not infallible. Occasionally, just as worker bees escape the parliament of the hive, so segregation distorters escape the majoritarian supervision of the parliament of genes. But usually, as Kropotkin hoped, the greater good prevails.

CHAPTER TWO

The Division of Labour

In which self-sufficiency proves to be much overrated

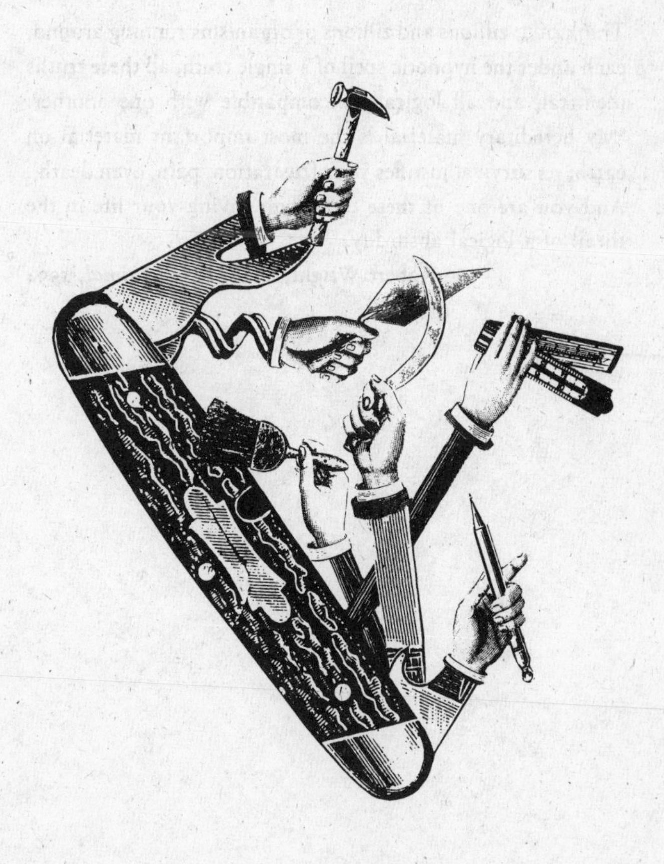

Think of it: zillions and zillions of organisms running around, each under the hypnotic spell of a single truth, all these truths identical, and all logically incompatible with one another: 'My hereditary material is the most important material on earth; its survival justifies your frustration, pain, even death'. And you are one of these organisms, living your life in the thrall of a logical absurdity.

Robert Wright, *The Moral Animal*, 1994

The Hutterites are more persistent and successful than most religious sects. Originating in Europe in the sixteenth century, they emigrated *en masse* to America in the nineteenth and founded agricultural communities all across North America. Their high birth rate, their general prosperity and their self-sufficiency, even on marginal land in Canada that other farmers have failed to cultivate, attest to a formula for living that is remarkably effective. That formula is, in a word, collectivism. Their principal virtue is *Gelassenheit*, which means, roughly, 'grateful acceptance of whatever God gives, even suffering and death, the forsaking of all self-will, all selfishness, all concern for private property'. 'True love,' said their leader, Ehrenpreis, in 1650, 'means growth for the whole organism, whose members are all interdependent and serve each other.'

In short, Hutterites are like bees: subservient parts of a greater whole. Indeed, they welcome the analogy and make it freely themselves. They have consciously recreated the same sorts of bulwarks against selfish mutiny that the coagulations of genes, cells and bees evolved millions of years ago. For instance, when a Hutterite community is large enough to split, it first prepares the new site for occupation, then matches people into pairs for age, sex and skills and only then, on the day of the split, do the people draw lots to decide who will occupy the new site and who will stay at the old one. There could not be a more precise analogy with the process of meiosis, the card-shuffling process which selects the lucky genes that will go into the egg and discards those that will not.[1]

The fact that such measures are necessary (and the harsh treatment meted out to Hutterites who exhibit egotism) testifies to the fact that subversive selfishness remains a persistent danger. In just the same way meiosis testifies to the ever-present possibility of genetic mutiny. This, argue some observers, far from demonstrating that Hutterites are human bees, proves the exact opposite. In commenting on David Wilson and Eliot Sober's analysis of the Hutterites, Lee Cronk argued: 'What the Hutterite example really demonstrates, is that it is very, very difficult to get people to act like Hutterites and most attempts to do so fail miserably.'

Yet all human beings share a fascinating taboo with the Hutterites, the taboo against selfishness. Selfishness is almost the definition of vice. Murder, theft, rape and fraud are considered crimes of great importance because they are selfish or spiteful acts that are committed for the benefit of the actor and the detriment of the victim. In contrast, virtue is, almost by definition, the greater good of the group. Those virtues (such as thrift and abstinence) that are not directly altruistic in their motivation are few and obscure. The conspicuously virtuous things we all praise – cooperation, altruism, generosity, sympathy, kindness, selflessness – are all unambiguously concerned with the welfare of others. This is not some parochial Western tradition. It is a bias shared by the whole species. Only something like glory, which is usually earned by selfish and sometimes violent acts, is an exception to this rule and it is an exception that proves the rule because glory is such an ambiguous virtue, shading so easily into vainglory.

My point is that we are all Hutterites at heart. Consciously or implicitly, we all share a belief in pursuing the greater good. We praise selflessness and decry selfishness. Kropotkin got it the wrong way round. The essential virtuousness of human beings is proved not by parallels in the animal kingdom, but by the very lack of convincing animal parallels. The thing that needs explaining about human beings is not their frequent vice, but their occasional virtue. George Williams put the question thus: 'How could maximizing selfishness produce an organism capable of often advocating, and occasionally practicing, charity towards strangers and even towards animals?'[2] The human obsession with virtue is unique to us and the truly social animals. Are we a coagulated species, too? Have we begun to lose our individuality to become parts of an overarching evolving thing called a society? Is that one of the things that are special about us? If so, we are odd in one crucial respect. We breed.

Although we have not surrendered reproduction to a queen, we human beings are surely as utterly dependent on each other as any ants or honey bees. As I write this, I am using software I did not invent on a computer I could never have made that depends on electricity I could not have discovered, and I am not worrying about where my next meal will come from because I know I can go and buy food from a shop. In a phrase, therefore, the advantage of society to me is the division of labour. It is specialization that makes human society greater than the sum of its parts.

Groupishness

If a creature puts the greater good ahead of its individual interests, it is because its fate is inextricably tied to that of the group: it shares the group's fate. A sterile ant's best hope of immortality is vicarious reproduction through the breeding of the queen, just as an aeroplane passenger's best hope of life is through the survival of the pilot. Vicarious reproduction through a relative explains how cells, corals and ants coagulate into teams of mostly harmonious collaborators. As we have seen, to enhance the selflessness of the individual cells, the embryo prevents their reproduction; to enhance the selflessness of worker ants, the queen renders them sterile.

Animal bodies, coral clones and ant colonies are just big families. Altruism within families is not a very surprising thing, because – we have seen – close genetic relatedness is a good reason for cooperation. Yet human beings cooperate at a level other than the family. Hutterite communities are not families. Nor are the bands of hunter-gatherer societies. Nor are the villages of farming people. Nor are armies, sports teams or religious congregations. To put the case the other way round: no known human society, with the possible exception of an abortive attempt by a West African kingdom in the nineteenth century, has ever even tried to restrict reproduction to one couple or even one polygamous man. So whatever human society is, it is not a big family. This makes its benevolent side much harder to explain. Indeed, human societies are conspicuous for their reproductive egalitarianism. Whereas many other group-living mammals – wolves, monkeys, apes – restrict the right to reproduce to a minority of males and sometimes of females as well, human beings all and everywhere expect to reproduce. 'However humans specialize and divide labor,' wrote Richard Alexander, 'they nearly always insist individually on the right to carry out all of the reproductive activities themselves.' The most harmonious societies, adds Alexander, are those that impose egalitarian reproductive opportunity on themselves: monogamous societies often prove more cohesive and better at conquest than polygamous ones, for example.[3]

Not only do people refuse to delegate the right to breed to others, they actually try to suppress kin favouritism for the greater good of society. Nepotism, after all, is a dirty word. Except in strictly private family matters, favouring your relatives over other members of the community is always a sign of corruption in any society. In his study of French villagers in the Jura in the early 1970s, Robert Layton found abundant evidence for the mistrust of nepotism. At the local level, certainly, people favoured their kin. But at the level of the commune, such favouritism was forcefully discouraged. The commune and the agricultural cooperative forbad fathers and sons or brothers seeking election at the same time. It was considered to be in everybody's interests to prevent the management of shared resources falling into the hands of kin-based factions. Nepotistic factions have a bad name in human societies: the Mafia is a prime example.[4]

This lack of nepotism makes the analogy between people and social insects faulty. Far from embracing vicarious reproduction, we seem to go to great lengths to avoid it. But it does not affect the analogy with chromosomes, which are even more egalitarian about reproduction. Chromosomes may not be altruistic – they do not surrender their right to replicate – but they are something other than selfish. They are 'groupish': they defend the integrity of the whole genome, suppressing selfish mutinies by individual genes.[5]

The parable of the pin-maker

There is one thing we have beaten the ants at: the division of labour. Ants do have divisions of labour – between workers and soldiers, nest workers and foragers, builders and hygiene specialists. But it is, by our standards, a pretty feeble division of labour. In ants there are four physically different castes of insect at most, yet there are often forty or more distinct tasks to be done. However, worker ants do change their duties as they age, multiplying the division of labour, and in some ants, such as army ants, the individuals work in teams, greatly expanding their skills.[6]

In honey bees there is no permanent division of labour at all, except that between the queen and the workers. Shakespeare's image in *Henry V* of bee magistrates, bee masons, bee porters and bee merchants is a fantasy. There are just workers, all of which are jacks of all trades. The advantages of society to a bee are that the colony is an efficient information-processing device for directing effort to where it is most rewarding. That does not require a division of labour.

By contrast, in human society, the advantages of society are those provided by the division of labour. Because each person is a specialist of some sort – usually from an early enough age to have become good at their chosen trade while still mentally malleable – the sums of all our efforts are greater than they would be if each of us had to be a jack of all trades. We recoil from only one specialization, the one that ants embrace most enthusiastically: the reproductive division of labour between breeders and helpers. In no human society do people routinely and enthusiastically surrender reproduction to their relatives. Maiden aunts and monks are nowhere numerous.

It is this synergy between specialists that makes human societies tick, and it is this that distinguishes us from all other social creatures. Only when we look at the society of cells that forms a body do we find a comparable complexity of specialized function. The division of labour is what makes a body worth inventing. A red blood cell is as valuable to a liver cell as vice versa. Between them they can achieve more than a single cell can ever manage. Each organ, each muscle, each tooth, each nerve and each bone plays its separate part in the whole enterprise. Nothing tries to do everything at once, which is why we can achieve rather more than slime moulds can. Indeed, at the very beginning of life itself, the division of labour was a crucial step. Not only did individual genes divide and share the functions of running a cell, but genes themselves had already specialized in storing information, dividing labour with proteins, which are specialized to carry out chemical and structural tasks. We know that this was a division of labour, because RNA, the more primitive and rarer of the material from which genes are made, is itself a jack of all trades, capable of both storing information and being a chemical catalyst. It is not as good at the former task as DNA, nor as good at the latter task as proteins.[7]

Adam Smith was the first to recognize that the division of labour is what makes human society more than the sum of its parts. In the opening chapter of his great book *An Inquiry into the Nature and Causes of the Wealth of Nations*, he chose to illustrate the point with the example of a pin-maker. Somebody not trained in pin-making could probably only make one pin a day, and even when practised he would only be able to make twenty or so. Yet, by dividing labour between pin-makers and non-pin-makers and by further dividing the task of pin manufacture between a number of specialist trades, we vastly increase the number of pins that can be made by each person. Ten people in a pin factory could and did, said Smith, produce 48,000 pins per day. To buy twenty pins from such a factory therefore costs only $^{1}/_{240}$ of a man-day, whereas it would have taken the purchaser a whole day at least to make them himself.

The reasons for this advantage, said Smith, lay in three chief consequences of the division of labour. By specializing in pin-making, the pin-maker improves his dexterity at pin-making through practice; he also saves the time that would otherwise be spent switching from task to task; and it pays him to invent, buy or use specialized machinery that speeds up the task. Writing at the dawn of the Industrial Revolution, Smith prophetically described in a few pages the sole reason why the material wealth of the country and the world would vastly increase in the ensuing two centuries and more. (He also recognized the alienating effects of too much specialization, writing that 'the man whose life is spent in performing a few simple operations . . . becomes as stupid and ignorant as it is possible for a human creature to become', thus presaging Marx and Charlie Chaplin.) Modern economists are unanimous in agreeing with Smith that the modern world owes its economic growth entirely to the cumulative effects of divisions of labour, as distributed by markets and fuelled by new technology.[8]

If biologists have not added to the theory proposed by Smith, they have at least tested it. Smith said two further things about the division of labour in society: that it increased with the size of the market, and that in a market of a given size it increased with improvement in transport and communication. Both maxims prove to be true of simple societies of cells, in this case a creature called *Volvox*, which lives in spheres of collaborating, but largely self-sufficient, cells. The bigger the *Volvox*, the more likely there is to be division of labour, some cells specializing in reproduction. And the more connection there is between the cells, the greater the division of labour. In *Merillisphaera*, the cells lose their private connections by which chemicals can flow from one to another, whereas in *Euvolvox*, such connections persist. *Euvolvox* can consequently pour more surplus effort into its specialized reproductive cells, which hence grow faster.[9]

John Bonner turned from studying the division of labour in slime moulds to bodies and societies. The facts prove Adam Smith right on the relation between scale and division of labour. Bigger bodies tend to have more different kinds of cells. Societies organized into bigger groups tend to have more different castes of occupation, ranging from the Tasmanians (now extinct), who lived in bands of fifteen and recognized only two castes, to the Maoris, who lived in groups of nearly 2,000 and recognized sixty different personal functions.[10]

Virtually nothing else of interest has been written about the division of labour since Adam Smith, either by biologists or by economists. In economics, only the conflict between division of labour and the inefficient monopolies it eventually creates has attracted much attention: if everybody is doing a different task, then nobody has the spur of competition.[11]

Biologists have been unable to explain why some ants have several worker castes and others just one. 'It seems odd,' wrote Michael Ghiselin, 'that biologists and economists alike have paid very little attention to the division of labor. Seeming to be too obvious to require explanation, it has been accepted as a mere brute fact, while its functional significance has been virtually ignored. Although labor is sometimes divided, sometimes combined, there are as yet no adequate explanations why.'[12]

Ghiselin discovered a paradox. Ants, termites and bees have in one sense become more specialized as they have abandoned 'hunter-gathering' for agriculture. Just like us, they use their divided-labour societies to grow crops or raise domesticated animals – in this case fungi and aphids rather than wheat and cattle, but the principle is much the same. On the other hand, social insects are far less specialized than solitary ones in their catholic tastes in food. Each beetle or butterfly larva eats just one kind of plant; each solitary wasp is superbly designed to kill just one kind of prey. But most ants eat almost anything that comes their way; honey bees visit flowers of all shapes and descriptions; termites eat wood, of whatever species of tree. Even the agriculturalists are generalists. Leaf cutter ants feed their fungi with leaves of many kinds of trees.

This is the great advantage of a division of labour: by *specializing* at the level of the individual, the species can *generalize* at the level of the colony. Hence the paradox that ants are far more numerous than beetles but far less diverse.[13]

Returning to Adam Smith's pin-maker, notice that both he and his customer are better off: the customer gets his pins cheaper, the pin-maker makes enough pins to exchange for a handsome supply of all the other goods he needs. From this followed perhaps the least appreciated insight in the whole history of ideas. Smith made the paradoxical argument that social benefits derive from individual vices. The cooperation and progress inherent in human society are the result not of benevolence, but of the pursuit of self-interest.

Selfish ambition leads to industry; resentment discourages aggression; vanity can be the cause of acts of kindness. In the most famous passage of his book, he wrote:

In almost every other race of animal each individual, when it is grown up to maturity, is entirely independent, and in its natural state has occasion for the assistance of no other living creature. But man has almost constant occasion for the help of his brethren, and it is in vain for him to expect it from their benevolence only. He will be more likely to prevail if he can interest their self-love in his favour, and show them that it is for their own advantage to do for him what he requires of them . . . It is not from the benevolence of the butcher, the brewer, or the baker, that we expect our dinner, but from their regard to their own interest. We address ourselves not to their humanity but to their self-love, and never talk to them of our own necessities but of their advantages. Nobody but a beggar chooses to depend chiefly on the benevolence of his fellow citizens.[14]

As Samuel Brittan has cautioned, Smith is easily misunderstood. The butcher may not be motivated by benevolence, but that does not mean he is motivated by callousness or a desire to be nasty to others. The pursuit of self-interest is as different from the pursuit of spite as it is from the pursuit of altruism.[15]

There is a beautiful parallel between what Smith meant and the human immune system. Our immune system depends on molecules that wrap themselves around foreign proteins. If they are to do so, the molecules must fit their targets exactly, and that means they are highly specific. Each antibody, or T cell, can attack only one kind of invader. Therefore, to work, the immune system must have almost countless types of defending cells. It has over a billion. Each one is rare, but is ready to multiply if it encounters its target. Its 'motive' is, in a sense, self-interested. When a T cell starts to multiply it is conscious of nothing and it is certainly not motivated by some urge to kill the invader. But it is, in a sense, driven by the need to multiply: the immune system is a competitive world in which only those cells thrive that divide when they get the chance. To multiply, a 'killer' T cell must get a supply of interleukins from a 'helper' T cell. The molecules that allow the 'killer' to obtain interleukins are the very same molecules that allow it to recognize invaders. And the 'helper' only helps because the molecule that compels it to help is the same molecule that it needs if it is to grow. So attacking the foreign invader is, for these cells, a by-product of the normal business of striving to grow and divide. The whole system is beautifully designed so that the self-interested ambitions of each cell can only be satisfied by the cell doing its duty for the body. Selfish ambitions are bent to the greater good of the body just as selfish individuals are bent by the market to the greater good of society. It is as if our blood were full of Boy Scouts running around looking for invaders because each time they found one they were rewarded with a chocolate.[16]

Smith's insight, translated into modern idiom, was that life is not a zero-sum game. A zero-sum game is one with a winner and a loser, like a tennis match. But not all games are zero-sum; sometimes both sides win, or lose. In the case of trade, Smith saw that because of the division of labour, my selfish ambition to profit from trading with you, and yours to profit from trading with me, can *both* be satisfied. We each act in self-interest, but we only benefit each other and the world. So, although Hobbes is right that we are vicious, not virtuous, Rousseau is right that harmony and progress are possible without government. The invisible hand guides us forward.

Such cynicism is shocking in a more self-conscious age. None the less, the subtle theme that good things can come of bad motives is one that cannot be ignored. It is an admission that good deeds are done, that the common good is to be had by humankind in society, but this does not require us to believe in angels. Self-seeking can produce benevolence. 'We are not ready to suspect any person of being defective in selfishness,' observed Smith in his *Theory of Moral Sentiments*. Indeed, Smith pointed out that benevolence is inadequate for the task of building cooperation in a large society, because we are irredeemably biased in our benevolence to relatives and close friends; a society built on benevolence would be riddled with nepotism. Between strangers, the invisible hand of the market, distributing selfish ambitions, is fairer.[17]

The technological Stone Age

Yet I have described the division of labour in modern society, not in the conditions of simple tribalism in which we spent most of our formative evolutionary aeons. Surely this division of labour is only recent? As Alfred Emerson, a termite expert influenced indirectly by Kropotkin, put it in 1960, 'As division of labor between specialists evolves, integration into higher unit systems also advances, and, as social homeostasis evolves, the individual human loses some portion of his self-regulation and becomes more dependent for his existence upon the division of labor and the integration of the social system.'[18]

Emerson was suggesting that the division of labour is something fairly new, something still progressing. Economists are even more apt to conclude that it is a modern invention. Back when everybody was a peasant everybody was a jack of all trades. Only since civilization spread its bounty among us have we begun to specialize.

I doubt this interpretation. I suspect that hunter-gatherers were subtly specialized hundreds of thousands of years ago. Modern hunter-gatherers certainly are: among the Ache of Paraguay, some men are known to be good at finding armadillos in their burrows; others are good at digging them out. Among Australian Aboriginals to this day there are people who are revered for certain skills and talents.[19]

When I was between eight and twelve, I attended a boarding school where, between the minor irritations of lessons and sports, the main activity was gang warfare. Like troops of chimpanzees we divided into gangs, each named after a leader, and set out to build impregnable fortresses in trees or underground tunnels from which to launch raids upon rival gangs. It seemed deadly serious at the time, though the casualties were slight. I vividly remember one day when, feeling confident and under-appreciated, I demanded to be allowed the privilege of being the gang member to climb a certain tree (why, I cannot remember). This was an act of breathtaking insubordination, because I was a junior member of the gang and everybody knew that X led all the tree climbing in our gang. I duly was allowed to fail in the task and X smugly resumed his appointed place in the hierarchy, while I dropped a few notches. We had a division of labour within the gang.

It is hard to imagine any group of grown men working together as a team for a fairly long period of time (as ancestral hunters would have done) without some similar sort of specializations emerging.

That this predates the Industrial Revolution is certain. In detailing the myriad different trades necessary for the creation of even the coarse woollen coat of a day labourer – the shepherds, weavers, merchants, toolmakers, carpenters, even the miners who mined the coal that powered the forge on which the shears were forged with which the shepherd clipped the wool – Adam Smith made plain the vast extent of the division of labour from which an eighteenth-century worker profited. Much the same could be said of medieval, Roman and Greek societies. Going farther back still, into the late Neolithic, the same argument applies. When the 5,000-year-old mummified corpse of a fully equipped Neolithic man turned up in a melting glacier high in the Tyrolean Alps in 1991, the variety and sophistication of his equipment was astonishing. Europe was in his day a tribal, thinly populated place of Stone Age culture. Copper was smelted but not yet bronze. Corn and cattle had long replaced hunting as the main livelihoods, but writing, law and government were unknown. Dressed in furs under a woven grass cloak, equipped with a stone dagger with an ash-wood handle, a copper axe, a yew-wood bow, a quiver and fourteen cornus-wood arrows, he also carried a tinder fungus for lighting fires, two birch-bark containers, one of which contained some embers of his most recent fire, insulated by maple leaves, a hazel-wood pannier, a bone awl, stone drills and scrapers, a lime-wood-and-antler *retoucheur* for fine stone sharpening, an antibiotic birch fungus as a medicine kit and various spare parts. His copper axe was cast and hammered sharp in a way that is extremely difficult to achieve even with modern metallurgical knowledge. It was fixed with millimetre precision into a yew haft that was shaped to obtain mechanically ideal ratios of leverage.

This was a technological age. People lived their lives steeped in technology. They knew how to work leather, wood, bark, fungi, copper, stone, bone and grass into weapons, clothes, ropes, pouches, needles, glues, containers and ornaments. Arguably, the unlucky mummy had more different kinds of equipment on him than the hiker couple who found him. Archaeologists believe he probably relied upon specialists for the manufacture of much of his equipment, and perhaps also for the tattoos that had been applied to his arthritic joints.[20]

Why stop there? I refuse to believe the same division of labour did not apply 100,000 years ago when our ancestors' bodies and brains were all but indistinguishable from ours. One man made stone tools, another knew how to find game, a third was especially good at throwing spears, a fourth could be relied upon as a strategist. Because of our tendency to imprint upon tasks that we are much exposed to during our youth, this division of labour would be reinforced by youth training. Thus, it is abundantly plain that the way to make a good tennis or chess player is first to find a young prodigy and then send him or her off to a school devoted to little else. I suspect the best hand-axe makers in the *Homo erectus* tribe started as apprentices to older men at a young age.

Men? I have ignored women in this fantasy not to slight them, but simply for purposes of illustrating the argument. Divisions of labour among women were probably as great as among men. However, there is one human division of labour that is extraordinarily marked in all known human societies: the division of labour between man and woman, or more especially between husband and wife. By gathering rare and protein-rich meat while his wife gathers plentiful but protein-poor fruits, the human couple gets the best of both worlds. No other primate exploits a sexual division of labour in this way (this is a subject I shall return to in Chapter Five).

The great advantage of human society is the division of labour, and the 'non-zero-sumness' it achieves. This phrase, invented by Robert Wright, neatly captures the point that society can be greater than the sum of its parts. But this still does not tell us how human society got started in the first place. We know it was not through nepotism. There is no evidence for the inbreeding and vicarious reproduction that is a necessary part of any nepotistic colony. So what was it? The strongest hypothesis is that it was reciprocity. In Adam Smith's words, 'the propensity to truck, barter and exchange one thing for another'.[21]

CHAPTER THREE

The Prisoner's Dilemma

In which computers learn to cooperate

I learn to do service to another, without bearing him any real kindness: because I foresee, that he will return my service, in expectation of another of the same kind, and in order to maintain the same correspondence of good offices with me or others. And accordingly, after I have serv'd him and he is in possession of the advantage arising from my action, he is induc'd to perform his part, as foreseeing the consequences of his refusal.

David Hume, *A Treatise of Human Nature*, 1740

In Puccini's opera *Tosca*, the heroine is faced with a terrible dilemma. Her lover Cavaradossi has been condemned to death by Scarpia, the police chief, but Scarpia has offered her a deal. If Tosca will sleep with him, he will save her lover's life by telling the firing squad to use blanks. Tosca decides to deceive Scarpia by agreeing to his request, but then stabbing him dead after he has given the order to use blanks. She does so, but too late discovers that Scarpia chose to deceive her too. The firing squad does not use blanks; Cavaradossi dies. Tosca commits suicide, and all three end up dead.

Though they did not put it this way, Tosca and Scarpia were playing a game, indeed the most famous game in all of game theory, an esoteric branch of mathematics that provides a strange bridge between biology and economics. The game has been central to one of the most exciting scientific discoveries of recent years: nothing less than an understanding of why people are nice to each other. Moreover, Tosca and Scarpia each played the game in the way that game theory predicts they should, despite the disastrous outcome for each. How can this be?

The game is known as the prisoner's dilemma, and it applies wherever there is a conflict between self-interest and the common good. Both Tosca and Scarpia would benefit if they stuck to their bargain: Tosca would save her lover's life and Scarpia would bed her. But as individuals each would benefit even more if he or she deceived the other into keeping his side of the bargain but did not keep his own: Tosca would save her lover and her virtue, whereas Scarpia would get lucky and be rid of his enemy.

The prisoner's dilemma presents us with a stark example of how to achieve cooperation among egoists – cooperation that is not dependent on taboo, moral constraint or ethical imperative. How can individuals be led by self-interest to serve a greater good? The game is called the prisoner's dilemma because the commonest anecdote to illustrate it describes two prisoners each faced with the choice of giving evidence against the other and so reducing his own sentence. The dilemma arises because if neither defects on the other, the police can convict them both only on a lesser charge, so both would be better off if they stayed silent, but each is individually better off if he defects.

Why? Forget prisoners and think of it as a simple mathematical game you play with another player for points. If you both cooperate ('stay silent') you each get three (this is called the 'reward'); if you both defect you each get one (the 'punishment'). But if one defects and the other cooperates, the cooperator gets nothing (the 'sucker's pay-off') and the defector gets five points (the 'temptation'). So, if your partner defects, you are better off defecting, too. That way you get one point rather than none. But if your partner cooperates, you are still better off defecting: you get five instead of three. *Whatever the other person does, you are better off defecting*. Yet, since he argues the same way, the certain outcome is mutual defection: one point each, when you could have had three each.

Do not get misled by your morality. The fact that you are both being noble in cooperating is entirely irrelevant to the question. What we are seeking is the logically 'best' action in a moral vacuum, not the 'right' thing to do. And that is to defect. It is rational to be selfish.

The prisoner's dilemma, broadly defined, is as old as the hills; Hobbes certainly understood it. So, too, did Rousseau, who in passing described a rather sophisticated version sometimes known as the co-ordination game in his famous but brief story of the stag hunt. Picturing a group of primitive men out hunting, he said:

> If it was a matter of hunting deer, everyone well realized that he must remain faithfully at his post; but if a hare happened to pass within reach

of one of them, we cannot doubt that he would have gone off in pursuit of it without scruple and, having caught his own prey, he would have cared very little about having caused his companions to lose theirs.[1]

To make it clear what Rousseau meant, suppose everybody in the tribe goes out to hunt a stag. They do so by forming a wide ring around the thicket in which the stag is lying, and walking inwards until the beast is finally forced to try to escape from the encircling cordon of hunters, at which point, if all goes well, it is killed by the closest hunter. But suppose one of the hunters encounters a hare. He can catch the hare for sure, but only by leaving the circle. That in turn leaves a gap through which the stag escapes. The hunter who caught the hare is all right – he has meat – but everybody else pays with an empty belly the price of his selfishness. The right decision for the individual is the wrong one for the group, so proving what a hopeless project social cooperation is (said misanthropic Rousseau bleakly).

A modern version of the stag hunt is the game suggested by Douglas Hofstadter called the 'wolf's dilemma'. Twenty people sit, each in a cubicle, with their fingers on buttons. Each person will get $1,000 after ten minutes, unless someone pushes his button, in which case the person who pushed the button will get $100 and everybody else will get nothing. If you are clever you do not push the button and collect $1,000, but if you are very clever, you see that there is a tiny chance that somebody will be stupid enough to push his or her button, in which case you are better off pushing yours first, and if you are very, very clever you see that the very clever people will deduce this and will push their buttons, so you, too, had better push yours. As in the prisoner's dilemma, true logic leads you into collective disaster.[2]

Old as the idea may be, the prisoner's dilemma was first formalized as a game in 1950 by Merril Flood and Melvin Dresher of the RAND corporation in California and first rephrased as an anecdote about prisoners by Albert Tucker of Princeton University a few months later. As Flood and Dresher realized, prisoners' dilemmas are all around us. Broadly speaking any situation in which you are tempted to do something, but know it would be a great mistake if everybody did the same thing, is likely to be a prisoner's dilemma. (The formal mathematical definition of the prisoner's dilemma is wherever the temptation is greater than the reward which is greater than the punishment which is greater than the sucker's pay-off, though the game changes if the temptation is huge.) If everybody could be trusted not to steal cars, cars need not be locked and much time and expense could be saved in insurance premiums, security devices and the like. We would all be better off. But in such a trusting world, an individual can make himself even better off by defecting from the social contract and stealing a car. Likewise, all fishermen would be better off if everybody exercised restraint and did not take too many fish, but if everybody is taking as much as he can, the fisherman who shows restraint only forfeits his share to somebody more selfish. So we all pay the collective price of individualism.

Tropical rain forests, bizarrely, are the products of prisoners' dilemmas. The trees that grow in them spend the great majority of their energy growing upwards towards the sky, rather than reproducing. If they could come to a pact with their competitors to outlaw all tree trunks and respect a maximum tree height of ten feet, every tree would be better off. But they cannot.

To reduce the complexity of life to a silly game is the kind of thing that gets economists a bad name. But the point is not to try to squeeze every real-life problem into a box called 'prisoner's dilemma', but to create an idealized version of what happens when collective and individual interests are in conflict. You can then experiment with the ideal until you discover something surprising and then return to the real world to see if it sheds light on what really happens.

Exactly this has occurred with the prisoner's dilemma game (although some theorists have to be dragged kicking and screaming back to the real world). In the 1960s, mathematicians embarked on an almost manic search for an escape from the bleak lesson of the prisoner's dilemma – that defection is the only rational approach. They repeatedly claimed to have found one, most notably in 1966 when Nigel Howard rephrased the game in terms of the players' intentions, rather than their actions. But Howard's resolution of the paradox, like every other one suggested, proved only to be wishful thinking. Given the starting conditions of the game, cooperation is illogical.

This conclusion was deeply disliked, not just because it seemed so immoral in its implications, but because it seemed so at odds with the way real people behave. Cooperation is a frequent feature of human society; trust is the very foundation of social and economic life. Is it irrational? Do we have to override our instincts to be nice to each other? Does crime pay? Are people honest only when it pays them to be so?

By the late 1970s, the prisoner's dilemma had come to represent all that was wrong with the economist's obsession with self-interest. If the game proved that the individually rational thing to do in such a dilemma was to be selfish, then that only proved the inadequacy of the assumption. Since people are not invariably selfish, then they must not be motivated by self-interest, but by the common good. Two hundred years of classical economics, built on the assumption of self-interest, was therefore barking up the wrong tree.

A brief digression on game theory: born, in 1944, in the fertile but inhuman brain of the great Hungarian genius Johnny von Neumann, it is a branch of mathematics that especially suits the needs of the 'dismal science' of economics. This is because game theory is concerned with that province of the world where the right thing to do depends on what other people do. The right way to add two and two does not depend on the circumstances, but the decision whether to buy or sell an investment does depend totally on the circumstances, and in particular on what other people decide. Even in that case, though, there may be a foolproof way to behave, a strategy that works whatever other people do. To find it in a real situation, like making an investment decision, is probably as close to impossible as makes no difference, but that does not mean the perfect strategy does not exist. The point of game theory is to find it in simplified versions of the world – to find the universal prescription. This became known in the trade as the Nash equilibrium, after the Princeton mathematician John Nash (who worked out the theory in 1951, and received the Nobel Prize for it in 1994 on recovering from a long schizophrenic illness). The definition of a Nash equilibrium is when each player's strategy is an optimal response to the strategies adopted by other players, and nobody has an incentive to deviate from their chosen strategy.

Consider, as an example, a game invented by Peter Hammerstein and Reinhard Selten. There are two individuals, called Konrad and Niko; they have to share money with each other. Konrad plays first and he must decide whether they will share the money equally (fair) or unequally (unfair). Niko plays second and he must decide how much money they will share: a high or a low amount. If Konrad plays unfair, he gets nine times as much as Niko. If Niko plays high, each gets ten times as much as he would under the low conditions. Konrad can demand nine times as much as Niko and there is nothing Niko can do about it. If he plays low, he punishes himself as well as Konrad. So he cannot even plausibly threaten to punish Konrad by playing low. The Nash equilibrium is for Konrad to play unfair and Niko to play high. This is not the ideal outcome for Niko, but it is the best of a bad job.[3]

Note that the best outcome is not necessarily achieved at the Nash equilibrium. Far from it. Often the Nash equilibrium lies with two strategies that deliver one or both partners into misery, yet neither can do any better by doing differently. The prisoner's dilemma is just such a game. When played a single time between naïve partners, there is only one Nash equilibrium: both partners defect.

Hawks and Doves

Then one experiment turned this conclusion on its head. For thirty years, it showed, entirely the wrong lesson had been drawn from the prisoner's dilemma. Selfishness was not the rational thing to do after all – so long as the game is played more than once.

Ironically, the resolution of this conundrum had been glimpsed at the very moment the game was invented, then subsequently forgotten. Flood and Dresher discovered a rather surprising phenomenon almost straight away. When they asked two colleagues – Armen Alchian and John Williams – to play the game 100 times for small sums of money, the guinea pigs proved surprisingly keen to cooperate: on sixty of the 100 trials both cooperated and captured the benefits of mutual aid. Each admitted in notes made throughout the game that he was trying to be nice to the other to lure him into being nice back – until the very end of the game, when each saw the chance for a quick killing at the other's expense. When the game was played repeatedly and indefinitely by a single pair of people, niceness, not nastiness, seemed to prevail.[4]

The Alchian–Williams tournament was forgotten, yet whenever people were asked to play the game, they proved remarkably likely to try cooperation, the logically wrong tactic. This undue readiness to cooperate was condescendingly put down to their irrationality and generally inexplicable niceness. 'Evidently,' wrote one pair of game theorists, 'the run-of-the-mill players are not strategically sophisticated enough to have figured out that strategy DD [both defect] is the only rationally defensible strategy.' We were too dense to get it right.[5]

In the early 1970s, a biologist rediscovered the Alchian–Williams lesson. John Maynard Smith, an engineer-geneticist, had never heard of the prisoner's dilemma. But he saw that biology could use game theory as profitably as economics. He argued that, just as rational individuals should adopt strategies like those predicted by game theory as the least worst in any circumstances, so natural selection should design animals to behave instinctively with similar strategies. In other words, the decision to choose the Nash equilibrium in a game could be reached both by conscious, rational deduction and by evolutionary history. Selection, not the individual, can also decide. Maynard Smith called an evolved instinct that met a Nash equilibrium an 'evolutionary stable strategy': no animal playing it would be worse off than an animal playing a different strategy.

Maynard Smith's first example was an attempt to shed light on why animals do not generally fight to the death. He set the game up as a contest between Hawk and Dove. Hawk, which is roughly equivalent to 'defect' in the prisoner's dilemma, easily beats Dove, but is bloodily wounded in a fight with another Hawk. Dove, which is equivalent to 'cooperate', reaps benefits when it meets another Dove, but cannot survive against Hawk. However, if the game is played over and over again, the softer qualities of Dove become more useful. In particular, Retaliator – a Dove that turns into a Hawk when it meets one – proves a successful strategy. We shall hear more of Retaliator shortly.[6]

Maynard Smith's games were ignored by economists, because they were in the world of biology. But in the late 1970s something rather disturbing began to happen. Computers started using their cold, hard, rational brains to play the prisoner's dilemma, and they began to do exactly the same thing as those foolish, naïve human beings – to be irrationally keen to cooperate. Alarm bells rang throughout mathematics. In 1979, a young political scientist, Robert Axelrod, set up a tournament to explore the logic of cooperation. He asked people to submit a computer program to play the game 200 times against each other program submitted, against itself and against a random program. At the end of this vast contest, each program would have scored a number of points.

Fourteen people submitted programs of varying complexity, and to general astonishment, the 'nice' programs did well. None of the eight best programs would initiate defection. Moreover, it was the nicest of all – and the simplest of all – that won. Anatol Rapoport, a Canadian political scientist with an interest in nuclear confrontation who was once a concert pianist and probably knew more about the prisoner's dilemma than anybody alive, submitted a program called Tit-for-tat, which simply began by cooperating and then did whatever the other guy did last time. Tit-for-tat is in practice another name for Maynard Smith's Retaliator.[7]

Alexrod held another tournament, asking people to try to beat Tit-for-tat. Sixty-two programs tried, and yet the one that succeeded was . . . Tit-for-tat itself! It again came out on top.

As Axelrod explained in his book on the subject:

What accounts for Tit-for-tat's robust success is its combination of being nice, retaliatory, forgiving and clear. Its niceness prevents it from getting into unnecessary trouble. Its retaliation discourages the other side from persisting whenever defection is tried. Its forgiveness helps restore mutual cooperation. And its clarity makes it intelligible to the other player, thereby eliciting long-term cooperation.[8]

Axelrod's next tournament pitted strategies against each other in a sort of survival-of-the-fittest war, one of the first examples of what has since become known as 'artificial life'. Natural selection, the driving force of evolution, is easily simulated on a computer: software creatures compete for space on the computer's screen in just the way that real creatures breed and compete for space in the real world. In Axelrod's version, the unsuccessful strategies gradually went to the wall, leaving the most robust program in charge of the field. This produced a fascinating series of events. First, the nasty strategies thrived at the expense of nice, naïve ones. Only retaliators like Tit-for-tat kept pace with them. But then, gradually, the nasty strategies ran out of easy victims and instead kept meeting each other; they too began to dwindle in numbers. Tit-for-tat now came to the fore and eventually once again, it stood in sole command of the battlefield.

Bat blood brothers

Axelrod thought his results might be of interest to biologists, so he contacted a colleague at the University of Michigan, none other than William Hamilton, who was immediately struck by a coincidence. More than ten years before, a young biology graduate student at Harvard named Robert Trivers had shown Hamilton an essay he had written. Trivers assumed that animals and people are usually driven by self-interest yet observed that they frequently cooperate. He argued that one reason self-interested individuals might cooperate was because of 'reciprocity': essentially, you scratch my back, and I'll scratch yours. A favour done by one animal could be repaid by a reverse favour later, to the advantage of both, so long as the cost of doing the favour was smaller than the benefit of receiving it. Therefore, far from being altruistic, social animals might be merely reciprocating selfishly desired favours. Encouraged by Hamilton, Trivers eventually published a paper setting out the argument for reciprocal altruism in the animal kingdom and citing some likely examples. Indeed, Trivers went as far as to describe the repeated prisoner's dilemma as a means of testing his idea and predicting that the longer a pair of individuals interacted, the greater the chance of cooperation. He virtually predicted Tit-for-tat.[9]

Here, suddenly, a decade later, in Hamilton's hands, was mathematical proof that Trivers's idea had real power. Axelrod and Hamilton published a joint paper called 'The evolution of cooperation', to draw biologists' attention to Tit-for-tat. The result was an explosion of interest and a search for real examples among animals.[10]

They were not long in coming. In 1983, the biologist Gerald Wilkinson returned to California from Costa Rica with a slightly grisly story of cooperation. Wilkinson had studied vampire bats, which spend the day in hollow trees and the night searching for large animals whose blood they can quietly sip from small cuts surreptitiously made in their skin. It is a precarious life, because a bat occasionally returns hungry, having either failed to find an animal or been prevented from drinking its fill from the wound. For old bats this happens only about one night in ten; but for young bats one night in three is unsuccessful, and two abortive nights in a row are not therefore uncommon. After as little as sixty hours without a blood meal, the bat is in danger of starving to death.

Luckily, however, for the bats, when they do get a meal they can usually drink more than they immediately need and the surplus can be donated to another bat by regurgitating some blood. This is a generous act, and the bats find themselves in a prisoner's dilemma: bats who feed each other are better off than bats that do not; however, bats that take food but do not give it are best off and bats that give food but do not receive it are worst off.

Since the bats tend to roost in the same places, and can live for a long time – up to eighteen years – they get to know each other as individuals, and they have the opportunity to play the game repeatedly, just like Axelrod's computer programs. They are not, incidentally, very closely related on average to their neighbouring roost-mates, so nepotism is not the explanation of their generosity. Wilkinson found that they seem to play Tit-for-tat. A bat that has donated blood in the past will receive blood from the previous donee; a bat that has refused blood will be refused blood in turn. Each bat seems to be quite good at keeping score, and this may be the purpose of the social grooming in which the bats indulge. The bats groom each other's fur, paying particular attention to the area around the stomach. It is hard for a bat that has a distended belly after a good meal to disguise the fact from another bat which grooms it. A bat that cheats is therefore soon detected. Reciprocity rules the roost.[11]

African vervet monkeys are similarly reciprocal. When played a tape recording of a call from one monkey requesting support in a fight, another monkey will respond much more readily if the caller has helped it in the past. But if the two are closely related, the second monkey's response does not depend so much on whether the first monkey has once helped it. Thus, as theory predicts, Tit-for-tat is a mechanism for generating cooperation between unrelated individuals. Babies take their mother's beneficence for granted and do not have to buy it with acts of kindness. Brothers and sisters do not feel the need to reciprocate every kind act. But unrelated individuals are acutely aware of social debts.[12]

The principal condition required for Tit-for-tat to work is a stable, repetitive relationship. The more casual and opportunistic the encounters between a pair of individuals, the less likely it is that Tit-for-tat will succeed in building cooperation. Trivers noticed that support for this idea can be found in an unusual feature of coral reefs: cleaning stations. These are specific locations on the reef where local large fish, including predators, know they can come and will be 'cleaned' of parasites by smaller fish and shrimps.

This form of cleaning is a vitally important part of being a tropical fish. More than forty-five species of fish and at least six of shrimp offer cleaning services on coral reefs, some of them relying on it as their sole source of food, and most of them exhibiting distinctive colours and activities that mark them out to potential clients as cleaners. Fish of all kinds visit them to be cleaned, often coming in from the open ocean, or out from hiding places under the reef, and some specially change their colour to indicate a need for a clean; it seems to be a particularly valuable service for large fish. Many fish spend as much time being cleaned as feeding, and return several times a day to be cleaned, especially if wounded or sick. If the cleaners are removed from a reef there is an immediate effect: the number of fish declines, and the number showing sores and infections increases rapidly as the parasites spread.

The smaller fish get food and the larger fish get cleaned: mutual benefit results. But the cleaners are often the same size and shape as the prey of the fish they clean, yet the cleaners dart in and out of the mouths of their clients, swim through their gills and generally dice with death. Not only are the cleaners unharmed, but the clients give careful and well understood signals when they have had enough and are about to move on; the cleaners react to these signals by leaving straight away. So strong are the instincts that govern cleaning behaviour that in one case cited by Trivers, a large grouper, reared for six years in an aquarium tank until he was four feet long, and accustomed to snapping up any fish thrown into his tank, reacted to the first cleaner he met by opening his mouth and gills to invite the cleaner in, even though he had no parasites at all.

The puzzle is why the clients do not have their cake and eat it: accept the cleaning services, but round off the session by eating the cleaner. This would be equivalent to defecting in the prisoner's dilemma. And it is prevented for exactly the same reason as defection is rare. The answer is roughly the same as an amoral New Yorker would probably give when asked why he bothers to pay his illegal-immigrant cleaning lady rather than just fire her and get another one next week: because good cleaners are hard to find. The client fish do not spare their cleaners out of a general sense of duty to future clients, but because a good cleaner is more valuable to them as a future cleaner than as a present meal. This is so only because the same cleaner can be found in the same spot on the same reef day after day for years on end. The permanence and duration of the relationship is vital to the equation. One-shot encounters encourage defection; frequent repetition encourages cooperation. There are no cleaning stations in the nomadic life of the open ocean.[13]

Another example Axelrod explored was the Western Front in the First World War. Because of the stalemate that developed, the war turned into one long battle over the same piece of ground, so that the encounters between any two units were repeated again and again. This repetition, like the repetition of games in the prisoner's dilemma, changed the sensible tactic from hostility to cooperation, and indeed the Western Front was 'plagued' by unofficial truces between Allied and German units that had been facing each other for some time. Elaborate systems of communication developed to agree terms, apologize for accidental infractions and ensure relative peace – all without the knowledge of the high commands on each side. The truces were policed by simple revenge. Raids and artillery barrages were used to punish the other side for defection, and these sometimes escalated out of control in just the way that blood feuds do. Thus, the situation bore a strong resemblance to Tit-for-tat: it produced mutual cooperation, but responded to defection with defection. The simple and effective 'remedy', put into practice by both sides' generals when the truces were discovered, was to move units about frequently, so that no regiment was opposite any other for long enough to build up a relationship of mutual cooperation.

However, there is a dark side to Tit-for-tat, as mention of the First World War reminds us. If two Tit-for-tat players meet each other and get off on the right foot, they cooperate indefinitely. But if one of them accidentally or unthinkingly defects, then a continuous series of mutual recriminations begins from which there is no escape. This, after all, is the meaning of the phrase 'tit-for-tat killing' in places where people are or have been addicted to factional feuding and revenge, such as Sicily, the Scottish borders in the sixteenth century, ancient Greece and modern Amazonia. Tit-for-tat, as we shall see, is no universal panacea.

But the lesson for human beings is that our frequent use of reciprocity in society may be an inevitable part of our natures: an instinct. We do not need to reason our way to the conclusion that 'one good turn deserves another', nor do we need to be taught it against our better judgements. It simply develops within us as we mature, an ineradicable predisposition, to be nurtured by teaching or not as the case may be. And why? Because natural selection has chosen it to enable us to get more from social living.

CHAPTER FOUR

Telling Hawks from Doves

In which developing a good reputation pays

Where it is in his own interest, every organism may reasonably be expected to aid his fellows. Where he has no alternative, he submits to the yoke of communal servitude. Yet given a full chance to act in his own interest, nothing but expediency will restrain him from brutalizing, from maiming, from murdering his brother, his mate, his parent, or his child. Scratch an 'altruist' and watch a 'hypocrite' bleed.

Michael Ghiselin, 1974, *The Economy of Nature and the Evolution of Sex*, University of California Press, Berkeley

For their size, vampire bats have very big brains. The reason is that the neocortex – the clever bit at the front of the brain – is disproportionately big compared to the routine bits towards the rear. Vampire bats have by far the largest neocortexes of all bats. It is no accident that they have more complex social relationships than most bats, including, as we have seen, bonds of reciprocity between unrelated neighbours in a group. To play the reciprocity game, they need to recognize each other, remember who repaid a favour and who did not, and bear the debt or the grudge accordingly. Throughout the two cleverest families of land-dwelling mammals, the primates and the carnivores, there is a tight correlation between brain size and social group. The bigger the society in which the individual lives, the bigger its neocortex relative to the rest of the brain. To thrive in a complex society, you need a big brain. To acquire a big brain, you need to live in a complex society. Whichever way the logic goes, the correlation is compelling.[1]

Indeed, so tight is the correlation that you can use it to predict the natural group size of a species whose group size is unknown. Human beings, this logic suggests, live in societies 150 strong. Although many towns and cities are bigger than this, the number is in fact about right. It is roughly the number of people in a typical hunter-gatherer band, the number in a typical religious commune, the number in the average address book, the number in an army company, the maximum number employers prefer in an easily run factory. It is, in short, the number of people we each know well.[2]

Reciprocity only works if people recognize each other. You cannot pay back a favour, or hold a grudge, if you do not know how to find and identify your benefactor or enemy. Moreover, there is one vital ingredient of reciprocity that our discussion of game theory has so far omitted: reputation. In a society of individuals that you recognize and know well, you need never play the prisoner's dilemma blindly. You can pick and choose your partners. You can pick those you know have cooperated in the past, you can pick those whom others have told you can be trusted, and you can pick those who signal that they will cooperate. You can discriminate.

Large, cosmopolitan cities are characterized by ruder people and more casual insult and violence than small towns or rural areas. Nobody would dream of driving in their home suburb or village as they do in Manhattan or central Paris – shaking fists at other drivers, hooting the horn, generally making clear their impatience. It is also widely acknowledged why this is the case. Big cities are anonymous places. You can be as rude as you like to strangers in New York, Paris or London and run only a minuscule risk of meeting the same people again (especially if you are in a car). What restrains you in your home suburb or village is the acute awareness of reciprocity. If you are rude to somebody, there is a good chance they will be in a position to be rude to you in turn. If you are nice to people, there is a good chance your consideration will be returned.

In the conditions in which human beings evolved, in small tribes where to meet a stranger must have been an extremely rare event, this sense of reciprocal obligation must have been palpable – it still is among rural people of all kinds. Perhaps Tit-for-tat is at the root of the human social instinct; perhaps it explains why, of all mammals, the human being has come closest to matching the naked mole rat in its social instincts.

The hunting of the snark

After Robert Axelrod's tournaments, there was a minor backlash against Tit-for-tat in game theory. Economists and zoologists alike began to crowd in with awkward objections.

The main problem that zoologists have with Tit-for-tat is that there are so few good examples of it from nature. Apart from Wilkinson's vampire bats, Trivers's reef cleaning stations and a handful of examples from dolphins, monkeys and apes, Tit-for-tat just is not practised. These few examples are a meagre return on the effort that went into looking for Tit-for-tat in the 1980s. To some zoologists the conclusion is stark: animals ought to play Tit-for-tat, but they don't.

A good example is lions. Lionesses live in tight-knit prides, each pride defending its territory against rival prides (male lions just attach themselves to prides for the sex, and do little of the work, either catching food or defending territory – unless it be from other males). Lionesses advertise their territorial ownership by roaring, so it is quite easy to fool them into thinking they face a serious invasion by playing tape-recorded roars in their territories. This Robert Heinsohn and Craig Packer did to some Tanzanian lions and watched their reaction.

The lionesses usually walk towards the sound to investigate, some rather enthusiastically, others a little reluctantly. This is fertile territory for Tit-for-tat. A brave lioness, who leads the approach to the 'intruder', should expect a reciprocal favour from a laggard, who hangs back: next time the laggard should lead, and risk danger. But Heinsohn and Packer found no such pattern. Leaders recognize laggards and keep looking back at them as if resentfully, but they usually lead the next time, too. Laggards are laggards.

> We suggest that female lions may be classified according to four discrete strategies: 'unconditional cooperators' who always lead the response, 'unconditional laggards' who always lag behind, 'conditional cooperators' who lag least when they are most needed, and 'conditional laggards' who lag farthest when they are most needed.[3]

There is absolutely no sign of punishment for the laggards, or reciprocity. The leaders just have to accept that their courage goes unappreciated. The lionesses do not play Tit-for-tat.

The fact that other animals do not often play Tit-for-tat does not prove that human beings do not build their societies upon reciprocity.

As we shall see in the next few chapters, the evidence that human society is riddled with reciprocal obligations is great and growing greater all the time. Like language and opposable thumbs, reciprocity might be one of those things that we have evolved for our own use, but that few other animals have found the use or the mental capacity for. Kropotkin may have been wrong, in other words, to expect mutual aid in insects just because it is present in people. None the less, the zoologists have a point. The simple idea of Tit-for-tat seems better suited to the simplified world of computer tournaments than the mess that is real life.

Tit-for-tat's Achilles' heel

Economists had a different problem with Tit-for-tat. Axelrod's discoveries, published in a series of papers and later in a book called *The Evolution of Co-operation*, caught the popular imagination and were widely publicized in the press. This fact alone would have earned them contempt from envious game theorists, and sure enough the sniping soon began.

Juan Carlos Martinez-Coll and Jack Hirshleifer put it bluntly: 'A rather astonishing claim has come to be widely accepted: to wit that the simple reciprocity behaviour known as Tit-for-tat is a best strategy not only in the particular environment modeled by Axelrod's simulations but quite generally.' They argued that one could just as easily design the conditions of a tournament in which Tit-for-tat would not do well, and, more worryingly, it seemed to be impossible to simulate a world where both nasty and nice strategies cohabited – yet that is the world we live in.[4]

Among the harshest critics has been Ken Binmore. He argues that it is vital to notice that, even in Axelrod's simulations, Tit-for-tat never wins a single game against a 'nastier' strategy: therefore, it is singularly bad advice to play Tit-for-tat if you enter a single game, rather than a series of games. You're just a sucker if you do. Axelrod, remember, added the scores obtained in matches between many different strategies. Tit-for-tat won by accumulating many high-scoring draws and losses, not by winning bouts.

Binmore believes that the very fact that we find Tit-for-tat such a natural idea – 'we all know deep down inside that it is reciprocity that keeps society going' – makes us uncritically keen to accept a mathematical rationalization of the notion. He adds: 'One must be very cautious indeed before allowing oneself to be persuaded to accept general conclusions extrapolated from computer simulations.'[5]

Much of this criticism misses the point. Axelrod should no more be criticized for failing to capture everything that happens in the world than Newton should be for failing to explain politics in terms of gravity. Everybody thought the prisoner's dilemma taught a bleak lesson, not only that it was rational to defect but also that it was stupid of people not to realize this. Yet Axelrod discovered that, in his words, 'the shadow of the future' alters this completely. A simple, nice strategy won his tournaments again and again. Even if his conditions later prove unrealistic, even if life is not precisely such a tournament, Axelrod's work has thoroughly demolished the working assumption of all those who had studied the subject before: that the *only* rational thing to do in a prisoner's dilemma is to be nasty. Nice guys can finish first.

As for the argument that Tit-for-tat wins by losing in high-scoring games, that is the whole point. Tit-for-tat loses or draws each battle but wins the war, by ensuring that most of its contests are high-scoring affairs, so it brings home the most points. Tit-for-tat does not envy or wish to 'beat' its opponent. Life, it believes, is not a zero-sum game: my success need not be at your expense; two can 'win' at once. Tit-for-tat treats each game as a deal struck between the participants, not a match between them.

Some of the highland people in central New Guinea, who live in a network of dangerous, unstable, but reciprocal alliances and feuds between tribes, have recently taken up football but, finding it a little too much for the blood pressure to lose a game, they have adjusted the rules. The game simply continues until each side has scored a certain number of goals. A good time is had by all, but there is no loser and every goal scorer can count themselves a winner. It is not a zero-sum game.

'Don't you see?' remonstrated the referee, a newly arrived priest, after one such drawn game. 'The object of the game is to try to beat the other team. Someone has to win!' The captains of the rival teams replied, patiently, 'No, Father. That's not the way of things. Not here in Asmat. If someone wins then someone else has to lose – and that would never do.'[6]

This is bizarre only because it is an idea we find so instinctively hard to grasp, at least in the context of games (I have my doubts about the joys of New Guinea football). Take the case of trade. It is axiomatic among economists that the gains from trade are mutual: if two countries increase their trade, both are better off. Yet this is not the way the man in the street, let alone his demagogue representative, sees it. To them, trade is a competitive matter: exports good, imports bad.

Imagine a football tournament slightly different from the New Guinea case. In this competition the winner of the league is the team to score the most goals, not the one that wins most games. Now imagine that some teams decide to play normal football, letting in as few goals as possible and scoring as many as possible. Other teams try a different strategy. They let the other team score a goal, then try to score themselves. If allowed, they return the favour; and so on. You can quickly see which teams will do best: the ones that are playing Tit-for-tat. Football has thus been changed from a zero-sum game to a non-zero-sum game. What Axelrod achieved was precisely to turn the prisoner's dilemma from a zero-sum game into a non-zero-sum game. Life is very rarely a zero-sum game.

However, in one important respect, Binmore and the other critics were right. Axelrod had been too hasty in concluding that Tit-for-tat itself is 'evolutionarily stable' – meaning that a population playing Tit-for-tat is immune to invasion by any other strategy. This conclusion was undermined by further computer-simulated tournaments, like Axelrod's third one, in which Rob Boyd and Jeffrey Lorberbaum showed that it was easy to design tournaments that Tit-for-tat does not win.

In these tournaments, to recapitulate, a random mix of strategies battle against each other for control of a finite space, by breeding at the rate defined by their points in the last game: 5, 3, 1 or 0. In these conditions, nasty strategies, such as 'Always defect', do well at first, exploiting the naïve cooperative strategies and crowding them out. But soon they get sluggish and feeble, because they only ever meet each other, and only ever get 1 point. Now is when Tit-for-tat comes into its own. Playing against 'Always defect', it soon defects to deprive the other of more than one 5-point temptation; but, playing against itself, it cooperates and reaps 3 points. Therefore, so long as one Tit-for-tat can find a few others and form even a small cooperative cluster, they can thrive and drive 'Always defect' extinct.[7]

But it is now that Tit-for-tat's weaknesses emerge. For example, Tit-for-tat is vulnerable to mistakes. Remember that it cooperates until it meets a defection, which it then punishes. When two Tit-for-tat players meet they cooperate happily, but if one starts to defect, purely by random mistake, then the other retaliates and before long both are locked in a miserably unprofitable round of mutual defections. To take an all-too-real example, when an IRA gunman in Northern Ireland, aiming at a British soldier, kills an innocent Protestant bystander, the mistake can spark a revenge murder of a randomly selected Catholic by a loyalist gunman, which in turn is avenged, and so on *ad infinitum*. Such a series of deaths in Northern Ireland was known for many years as tit-for-tat killing.

Because of such weaknesses, it was apparent that Tit-for-tat's success in the Axelrod tournaments was largely a function of their form. The tournaments just happened not to show up these weaknesses. In a world where mistakes are made, Tit-for-tat is a second-rate strategy, and all sorts of other strategies prove better. The clear conclusions that Axelrod had drawn became clouded as ever more rococo elaborations of new strategies were invented.

Enter Pavlov

The scene now shifts to Vienna, where Karl Sigmund, an ingenious mathematician with a playful cast of mind, was giving a seminar on game theory to a group of students one day in the late 1980s. One of the students in the audience, Martin Nowak, decided there and then to abandon his own studies of chemistry and become a game theorist. Sigmund, impressed by Nowak's determination, set him the task of solving the thicket of complication that had entrapped the prisoner's dilemma in the wake of Tit-for-tat. Find me the perfect strategy in a realistic world, said Sigmund.

Nowak designed a different kind of tournament, one in which nothing was certain, and everything was statistically driven. Strategies made random mistakes with certain probabilities, or switched between tactics in the same probability-driven manner. But the system could 'learn' or evolve by keeping improvements and dropping unsuccessful tactics. Even the probabilities with which they did things were open to gradual evolutionary change. This new realism proved remarkably helpful, stripping away all the rococo complications. Instead of several strategies equally capable of winning the game, one clearly came out on top. It was not Tit-for-tat but a very near relation called Generous-Tit-for-tat (which I will call Generous, for short).

Generous occasionally forgives single mistakes. That is, about one-third of the time it magnanimously overlooks a single defection. To forgive all single defections – a strategy known as Tit-for-two-tats – is merely to invite exploitation. But to do so randomly with a probability of about a third is remarkably effective at breaking cycles of mutual recrimination while still remaining immune to exploitation by defectors. Generous will spread at the expense of Tit-for-tat in a computer population of pure Tit-for-tat players that are making occasional mistakes. So, ironically, Tit-for-tat merely paves the way for a nicer strategy than itself. It is John the Baptist, not the Messiah.

But neither is Generous the Messiah. It is so generous that it allows even nicer, more naïve strategies to spread. For example, the simple strategy 'Always cooperate' can thrive among Generous players, though it does not actually defeat them; it can creep back from the dead. But 'Always cooperate' is a fatally generous strategy and is easily invaded by 'Always defect', the nastiest strategy of all. Among Generous players, 'Always defect' gets nowhere; but when some start playing 'Always cooperate', it strikes. So, far from ending up with a happy world of reciprocity, Tit-for-tat ushers in Generous, which can usher in 'Always cooperate', which can unleash perpetual defection, which is back where we started from. One of Axelrod's conclusions was wrong: there is no stable conclusion to the game.

As the summer of 1992 began, Sigmund and Nowak were depressed by their conclusion that there is no stable solution to the prisoner's dilemma game. It is the sort of untidy decision game theorists dislike. But, as luck would have it, Sigmund's wife, a historian, was due to spend the summer in Schloss Rosenburg, a fairytale castle in the Waldviertel of lower Austria, as the guest of a Graf whose ancestry she was studying. Sigmund asked Nowak along and they brought a pair of laptop computers to play prisoner's dilemma tournaments. The castle is used as a falconry school and, by day, the two mathematicians found themselves distracted every two hours by the thousand-foot dives of imperial eagles practising their technique over the castle courtyard. It was a suitably medieval setting for the jousting matches they organized inside their computers.

They went back to the beginning and entered into the lists of their tournaments all sorts of strategies that had been rejected before, trying to find one that not only won, but could remain stable after winning the tournament. They tried giving their playing automata a slightly better memory. Instead of just reacting to the partner's last play, as Tit-for-tat does, the new strategies remembered their own last play as well and acted accordingly. One day, quite suddenly, as the eagles dived past the window, inspiration struck. An old strategy first tried by – who else? – Anatol Rapoport, suddenly kept coming out on top. Rapoport had dismissed the strategy as hopeless, calling it Simpleton. But that was because he had pitted it against 'Always defect', against which it was indeed naïve. Nowak and Sigmund entered it into a world dominated by Tit-for-tat and it not only defeated the old pro, but proved invincible thereafter. So, although Simpleton cannot beat 'Always defect', it can steal the show once Tit-for-tat has extinguished 'Always defect'. Once again, Tit-for-tat plays John the Baptist.

Simpleton's other name is Pavlov, though some say this is even more misleading – it is the opposite of reflexive. Nowak admits that he should call it by the cumbersome but accurate name of Win-stay/Lose-shift, but he cannot bring himself to do so, so Pavlov it remains. Pavlov is like a rather simplistic roulette gambler. If he wins on red, he sticks to red next time; if he loses, he tries black next time. For win, read 3 or 5 (reward and temptation); for lose, read 1 or 0 (punishment and sucker's pay-off). This principle – that you don't mend your behaviour unless it is broken – underlies a lot of everyday activities, including dog training and child-rearing. We bring up our children on the assumption that they will continue doing things that are rewarded and stop doing things that are punished.

Pavlov is nice, like Tit-for-tat, in that it establishes cooperation, reciprocating in that it tends to repay its partners in kind, and forgiving, like Generous, in that it punishes mistakes but then returns to cooperating. Yet it has a vindictive streak that enables it to exploit naïve cooperators like 'Always cooperate'. If it comes up against a sucker, it keeps on defecting. Thus it creates a cooperative world, but does not allow that world to decay into a too-trusting Utopia where free-riders can flourish.

Yet Pavlov's weakness was well known. As Rapoport had discovered, it is usually helpless in the face of 'Always defect', the nasty strategy. It keeps shifting to cooperation and getting the sucker's pay-off – hence its original name of Simpleton. So Pavlov cannot spread until Tit-for-tat has done its job and cleared out the bad guys. Nowak and Sigmund, however, discovered that Pavlov only shows this flaw in a deterministic game – one in which all the strategies are defined in advance. In their more realistic world of probability and learning, where each strategy rolled a die to decide what to do next, something very different happened. Pavlov quickly adjusted its probabilities to the point where its supremacy could not be challenged by 'Always defect'. It was truly evolutionarily stable.[8]

The fish that play chicken

Do animals or people use Pavlov? Until Nowak and Sigmund published their idea, one of the neatest examples of Tit-for-tat from animals was an experiment by Manfred Milinski using fish called sticklebacks. Sticklebacks and minnows are eaten by pike, and they react to the presence of a pike by leaving the school in a small scouting party and approaching it cautiously to assess the danger it poses. This apparently foolish courage must have some reward; naturalists think it gives the prey some valuable information. If, for example, they conclude that the pike is not hungry or has just fed, they can return to feeding themselves.

When two sticklebacks inspect a predator together, they move forward in a series of short spurts, one fish taking the initiative and risk each time. If the pike moves, both dash back again. Milinski argued that this was a series of small prisoner's dilemmas, each fish having to offer the 'cooperative' gesture of the next move forward, or take the 'defector's' option of letting the other fish go ahead alone. By an ingenious use of mirrors, Milinski presented each fish with an apparent companion (in fact its own reflection) that either kept up with it or lagged further and further behind as it got nearer the pike. Milinski at first interpreted his results in terms of Tit-for-tat: the trial fish was bolder with a cooperator than a defector. But, on hearing about Pavlov, he recalled that his fish would seem to switch back and forth between cooperation and defection when presented with a consistently defecting companion that had previously once cooperated – like Pavlov but unlike Tit-for-tat.

It may seem absurd to look at fish, expecting to find sophisticated game theorists, but there is, in fact, no requirement in the theory that the fish understand what it is doing. Reciprocity can evolve in an entirely unconscious automaton, provided it interacts repeatedly with other automata in a situation that resembles a prisoner's dilemma – as the computer simulations prove. Working out the strategy is the job not of the fish itself, but of evolution, which can then program it into the fish.

Pavlov is not the end of the story. Since Nowak has moved to Oxford it became inevitable that somebody at Cambridge had to take up the challenge of surpassing Pavlov. That somebody was Marcus Frean, who tried a new trick of playing the game in a more realistic fashion, in which the two players do not have to move simultaneously. Vampire bats do not do each other favours at the same moment. They take turns – there would be no point in simply swapping food for fun. Frean ran a tournament of this 'alternating prisoner's dilemma' inside his computer and, sure enough, there evolved a strategy that defeated Pavlov. Frean calls it Firm-but-fair. Like Pavlov it cooperates with cooperators, returns to cooperating after a mutual defection and punishes a sucker by further defection. But unlike Pavlov it continues to cooperate after being the sucker in the previous round. It is, therefore, slightly nicer.

The significance of this is not to raise Firm-but-fair into a new god, but to notice that making the game asynchronous makes guarded generosity even more rewarding. This accords with common sense. If you have to act before your partner and vice versa, it pays to try to elicit cooperation by being nice. You do not, in other words, greet strangers with a scowl lest they form a bad opinion of you; you greet them with a smile.

The first moralizers

Yet a more formidable problem looms. The prisoner's dilemma is a two-person game. Cooperation can, it seems, evolve spontaneously if a pair of individuals plays the game together indefinitely. Or, to put it more accurately, in a world where you only ever meet your immediate neighbour, it pays to be nice to him. But the world is not like that.

Reciprocity has a hard enough time producing cooperation even within a pair: the pair must be able to police their contract by being sure of encountering and recognizing each other again. How much harder is it among three individuals or more? The larger the group, the more inaccessible are the benefits of cooperation and the greater the obstacles that stand in the way. Indeed, Rob Boyd, a theorist, has argued that not only Tit-for-tat but any reciprocal strategy is simply inadequate to the task of explaining cooperation in large groups. The reason is that a successful strategy in a large group must be highly intolerant of even rare defection, or else free-riders – individuals who defect and do not reciprocate – will rapidly spread at the expense of better citizens. But the very features that make a strategy intolerant of rare defection are those that make it hard for reciprocators to get together when rare in the first place.[9]

Boyd himself provides one answer. Reciprocal cooperation might evolve, he suggests, if there is a mechanism to punish not just defectors, but also those who fail to punish defectors. Boyd calls this a 'moralistic' strategy, and it can cause any individually costly behaviour, not just cooperation, to spread, whether it causes group benefit or not. This is actually a rather spooky and authoritarian message. Whereas Tit-for-tat suggested the spread of nice behaviour among selfish egoists without any authority to tell them to be nice, in Boyd's moralism we glimpse the power that a fascist or a cult leader can wield.

There is another and potentially more powerful answer to the problem of free-riders in large groups: the power of social ostracism. If people can recognize defectors, they can simply refuse to play games with them. That effectively deprives the defectors of Temptation (5), Reward (3) and even Punishment (1). They do not get a chance to accumulate any points at all.

Philip Kitcher, a philosopher, designed an 'optional prisoner's dilemma' game to explore the power of ostracism. He populated a computer with four kinds of strategist: discriminating altruists, who play only with those who have never defected on them before; willing defectors, who always try defecting; solitaires, who always opt out of any encounter; and selective defectors who are prepared to play with those who have never defected before – but then, treacherously, defect on them.

Discriminating altruists (DAs) invading a population of solitaires soon prevail, because they find each other and reap the Reward. But surprisingly, selective defectors cannot then invade a population of

DAs, whereas DAs can invade one of selective defectors. In other words, discriminating altruism, which is just as 'nice' as Tit-for-tat, can reinvade anti-social populations. It is no more stable than Tit-for-tat, because of a similar vulnerability to a gradual take-over by undiscriminating cooperators. But its success hints at the power of ostracism to help in solving prisoners' dilemmas.[10]

Kitcher's programs relied entirely on the past behaviour of partners to judge whether they could be trusted. But discriminating between potential altruists need not be so retrospective. It might be possible to recognize and avoid potential defectors in advance. Robert Frank, an economist, set up an experiment to find out. He put a group of strangers in a room together for just half an hour, and asked them each to predict privately which of their fellow subjects would cooperate and which would defect in a single prisoner's dilemma game. They proved substantially better than chance at doing so. They could tell, even after just thirty minutes' acquaintance, enough about somebody to predict his cooperativeness.

Frank does not claim that this is too surprising. We spend a good deal of our lives assessing the trustworthiness of others, and we make instant judgements with some confidence. He poses a thought experiment for those unconvinced. 'Among those you know (but have never observed with respect to pesticide disposal), can you think of anyone who would drive, say, forty-five minutes to dispose of a highly toxic pesticide properly? If yes, then you accept the premise that people can predict cooperative predispositions.'[11]

Can fish be trusted?

Now, suddenly, there is a new and powerful reason to be nice: to persuade people to play with you. The reward of cooperation, *and the temptation of defection*, are forbidden to those who do not demonstrate trustworthiness and build a reputation for it. Cooperators can seek out cooperators.

Of course, for such a system to work, individuals must learn to recognize each other, which is not an easy feat. I have no idea whether a herring in a shoal of 10,000 fish or an ant in a colony of 10,000 insects, ever says to itself: 'There's old Fred again.' But I feel quite safe in assuming that it does not. On the other hand I feel equally secure in asserting that a vervet monkey probably knows by sound and sight every other member of its troop, because the primatologists Dorothy Cheney and Robert Seyfarth have proved as much. Therefore, a monkey has the necessary attributes for reciprocating cooperation, but a herring does not.

However, I may be maligning fish. Manfred Milinksi and Lee Alan Dugatkin have discovered a remarkably clear pattern of ostracism in stickleback fish when they risk their lives to inspect predators. A fish will tolerate more defection on the part of another fish that has continuously cooperated in the past than one that has not cooperated. And sticklebacks tend to pick the same partners to accompany them on predator-inspection visits each time – choosing partners who are consistently good cooperators. In other words, not only are the sticklebacks quite capable of recognizing individuals, but they seem capable of keeping individual scores – remembering which fish can be 'trusted'.

This is a puzzling discovery, in the light of how rare reciprocal cooperation is in the animal kingdom. Compared to nepotism, which accounts for the cooperation of ants and every creature that cares for its young, reciprocity has proved to be scarce. This, presumably, is due to the fact that recriprocity requires not only repetitive interactions, but also the ability to recognize other individuals and keep score. Only the higher mammals – apes, dolphins, elephants and a few others – are thought to possess sufficient brain power to be so discriminating for more than a handful of individuals. Now we know that sticklebacks can also keep score, at least for one or two 'friends', this assumption may have to be relaxed.

Whatever the capability of sticklebacks, there is no doubt that human beings, with their astonishing ability to recall the features of even the most casual acquaintance and their long lives and long memories, are equipped to play optional prisoner's dilemma games with far greater aplomb than any other species. Of all the species on the planet most likely to satisfy the criteria of prisoner's dilemma tournaments – the ability to 'meet repeatedly, recognize each other and remember the outcomes of past encounters', as Nowak has put it – human beings are the most obvious. Indeed, it might be what is special about us: we are uniquely good at reciprocal altruism.

Think about it: reciprocity hangs, like a sword of Damocles, over every human head. He's only asking me to his party so I'll give his book a good review. They've been to dinner twice and never asked us back once. After all I did for him, how could he do that to me? If you do this for me, I promise I'll make it up later. What did I do to deserve that? You owe it to me. Obligation; debt; favour; bargain; contract; exchange; deal . . . Our language and our lives are permeated with ideas of reciprocity. In no sphere is this more true than in our attitude to food.

CHAPTER FIVE

Duty and the Feast

In which human generosity with food is explained

He who understands baboon would do more towards metaphysics than Locke.

Charles Darwin, *Notebooks*[1]

Imagine if sex were an activity normally carried out communally and publicly, but eating was something done secretly and privately. There is no particular reason why the world could not be organized that way, so that it seemed positively odd to want to have sex alone and rather shameful to be caught eating in public. No reason except human nature. It is simply part of our make-up that food is communal and sex is private. It is so deeply ingrained in the human mind that the reverse is unthinkably weird. The bizarre notion, beloved of various historians, that sexual privacy was a cultural invention of medieval Christendom, has long since been exploded. All over the world, whatever god people worship, and however many or few clothes they wear in public, sex is a secret act to be done quietly when everybody else is asleep or out in the fields in the daytime where nobody else can see. It is a universal human characteristic. Eating food, on the other hand, is just as universally a communal activity.[2]

Throughout the world people gather together to feed. To eat in a group is normal and expected. We gather round the table for dinner, we meet a friend at a restaurant to share a meal, we join colleagues over sandwiches for working lunches, we woo and are wooed over candle-lit food. If a stranger is invited into your house or office she is offered food – even if it is only coffee and biscuits. To eat is to share. To offer to share food is simply a social instinct.

The food that we share most is meat. The larger and more social the meal the more unthinkable it is that meat would not be part of it. A description of a Roman or medieval feast is simply a list of meats: larks and boars, capons and beef. No doubt there were vegetables, too, but what made a feast different from a normal meal was the quantity of meat. Or perhaps the chronicler just found the meat more worthy of mention than the parsnips. Meat still occupies this feastly role. You would think it odd if you attended a glittering banquet given by a wealthy company at a four-star hotel and were served a main course of pasta, but you would think nothing of having just pasta as a main course when eating at home.

Even in the home, meat is still seen as the central ingredient of most meals. What's for dinner? asks somebody. Steak, replies whoever cooked it, or fish, omitting to mention the potatoes and cabbage that will be – nutritionally – just as important a part of the meal. Meat is usually placed on the plate first, or most centrally. The man, the head of the house, used to have the ceremonial role of carving it up, equitably, in front of the assembled guests – he still does in some households. How many of the snacks you graze during the day consist of meat? Very few.[3]

I have taken these examples from a narrow cultural perspective and described some parochial Western habits. But I maintain that much the same holds true throughout our species in all cultures and all continents: that eating is largely communal, social and shared; and that meat is usually, though not always, the most communal and shared of all foods. The most fundamentally selfless and communitarian thing we do is to share food; it is the very basis of society. Sex we do not share; we are possessive, jealous and secretive, prone to murdering our sexual rivals and guarding our partners if given the chance. But food is something to share.

Food sharing is, if not a uniquely human trait, then at least a peculiarity of our species, apparent even in small children. Birute Galdikas, who studies orang-utans in the forests of Borneo, brought up her own child, Binti, in a camp full of baby apes. This enabled her to notice what people would normally take for granted, a sharp difference between human beings and orang-utans in their attitude to food sharing. 'Sharing food seemed to give [Binti] great pleasure. In contrast, Princess, like any orang-utan, would beg, steal, and gobble food at every opportunity. Sharing food was not part of her orang-utan nature at that age.'[4]

How many other things that you possess are you prepared to share in the way that you are prepared to share food? We seem to have stumbled here upon a curiously generous aspect of human nature, a strange source of benevolence that people simply do not show with respect to other possessions. In the battle to capture the benefits of virtue – the division of labour and the opportunities for cooperative synergy – it was hunting for meat that granted our species its first great opportunity.

Meat for sex in chimpanzees

Anthropologists have long recognized that food sharing is a universal human habit and that meat is shared more than other food. This is principally because meat tends to come in larger packages than other food. The Yanomamo of Venezuela share large game killed in the forest but not small game or plantains grown in the band's gardens. Among the Ache of Paraguay a hunter gives away ninety per cent of the meat from a monkey or a peccary (a wild pig), but much less of a palm tree pith or a small armadillo. Among the Tiwi of Arnhem Land in Australia, the family of the hunter keeps eighty per cent of the smallest game, but only twenty per cent of the meat from animals larger than twelve kilograms.

We human beings are the most carnivorous of all primates. Judging even by the relatively vegetarian standards of most modern hunter-gatherers, rather than by the excessively meat-dependent habits of affluent Westerners, we still eat far more meat than our nearest rivals, the baboons and chimpanzees. The !Kung of the Kalahari, for example, eat a diet consisting of roughly twenty per cent meat, whereas in the diet of the chimpanzees of Tanzania, meat comprises at most five per cent of the food eaten (by weight). However, this is not to deny the importance to chimpanzees of meat. They put inordinate effort into hunting, and rarely pass up a good chance of getting meat. Baboons, likewise, clearly consider the meat of gazelle fawns to be a special treat.

But even among chimpanzees we can glimpse signs of cooperative culture that meat eating seems to induce. Hunting meat is a social activity in chimps, carried out mostly by parties of males. The bigger the hunting party, the bigger the success rate. At Gombe in Tanzania, the principal prey of the chimps is the red colobus monkey, and overall they are successful in about half the hunts they undertake, though the success rate can rise to nearly 100 per cent if more than ten males are in the party. The chimps usually catch a baby colobus, a small prize that, if shared among a large group of adult chimps, does not provide a large meal.

So why do they hunt at all? For a time scientists worried that hunting might be aberrant behaviour caused by the presence of human observers following the chimpanzees and frightening the monkeys in ways that made them easier to catch. But hunting has since been seen among chimps elsewhere, and it continued at Gombe even during the years that scientists were not present, so they now accept that it is normal. A curious theory is now emerging among scientists who have studied the behaviour in the wild. The chimps, they believe, are not hunting for nutritional reasons at all, but for social and reproductive reasons. They hunt in order to have sex.

If a party of chimpanzees comes across a troop of colobus monkeys in the forest, they sometimes choose to hunt them and sometimes do not. If the chimp party is large, they are more likely to start a hunt, which makes sense because they are also more likely to succeed. But by far the most reliable predictor of whether the chimps will hunt is the presence or absence of sexually receptive females in the party. If one of the females in the party is a 'swollen' female – with the sexual swelling that indicates oestrus – then the males in the party will usually start a hunt. Once they have caught a monkey they will preferentially give some of it to the swollen female. And, surprise, surprise, the female proves more likely to have sex with the males that are more generous with meat.

This is a common habit among scorpion flies: the male brings a large bribe, such as a dead insect prey, and feeds it to the female, who then allows him to mate with her. The bargain is not quite so blatant in chimpanzees, but it is there none the less. Food is shared by the males with receptive females in exchange for sex.[5]

The sexual division of labour

Chimpanzees are our closest relatives. Most anthropologists believe that the first proto-humans – the Australopithecines – lived in societies rather like those of chimps, with many adult males sharing and competing for many adult females. There is no good evidence for this except that no ground-dwelling monkey or ape living on the savanna has any other social system.

So let us assume for the moment that human hunting started for the same reasons as for chimpanzees. Proto-men went hunting for meat to offer to proto-women in exchange for sex. It is not all that unreasonable an assumption, and something rather like this occurs in Henry Fielding's great novel *Tom Jones*, in which meat and sex are closely juxtaposed. Indeed, in modern hunter-gatherers, it is uncomfortably close to the truth. In those tribes where promiscuity is common, men spend more time hunting for meat.

Take two examples. The Ache are a tribe with relative sexual freedom. Women are free to meet men other than their husbands, extra-marital affairs are fairly common, flirtatious talk is allowed and different bands often meet. Promiscuity is not encouraged or approved, but it is certainly possible. Ache men are keen hunters, spending on average seven hours a day in the forest in search of prey. Successful hunters have more affairs. The Hiwi, by contrast, are puritans. They have a male-biased sex ratio, they do not like to visit other bands and there is virtually no extra-marital sex. The Hiwi men have just as much spare time as the Ache, but they spend little of it hunting: a day or two a week for a couple of hours at a time. The meat they get goes to their families. In Africa a similar contrast is found in the Hadza and the !Kung. The Hadza men are obsessive hunters and promiscuous seducers. The !Kung are intermittent hunters and largely faithful husbands.[6]

Four cases do not make a theory, but it seems plausible to suppose that there lingers in the modern male mind a predisposition to respond to mating opportunities by trying to hunt for meat. Yet there is much more to human hunting than that. After all, meat is a staple food in many foraging people; it is not a rare luxury. The meat-seduction pattern may have been the origin of food sharing in human beings, but it has evolved into something much more fundamental and crucial, an economic institution that is a vital part of all human societies: the sexual division of labour.

There is one big difference between human beings and chimpanzees and that is the institution we call marriage. In virtually all human cultures, including hunter-gatherer societies, males monopolize their mates, and vice versa. Even if he ends up with more than one wife (as a few men do in hunter-gatherer bands), each man enters into a long-term relationship with each woman who bears his children. Unlike a male chimpanzee who mostly loses interest in a female as soon as she is no longer in oestrus, the man remains in close and jealous sexual union with his wife for many years, if not the rest of his life. Long-term pair bonds are not a cultural construct of our particular society; they are a habit universal to our species.[7]

As a consequence, there is a different motive for male hunting. A man can go hunting to get food for his children, just as a male hawk or a fox does. This only increases the advantage of the man hunting for meat. Living as he does in a pair bond, the man can share all his meat with his wife who can share all her vegetables with him. Both are better off. The division of labour is born; each half of the trading pair is better off than it would be on its own. The woman can gather enough roots, berries, fruits and nuts for two while the man catches a pig or a rabbit that gives the stew a rich mix of proteins and vitamins.

Forty years ago, anthropologists noticed that a sexual division of labour was a part of virtually all human societies. In the 1960s, squeamish at the sexism this implied, they dropped the subject and blamed the differences on patriarchal prejudice. But that explanation will not wash. A sexual division of labour is not a symptom of prejudice. It occurs in the most egalitarian societies. Anthropologists are virtually unanimous in agreeing that hunter-gatherers are less sexist than farming people, and women are less dominated. But they are equally unanimous in noticing an allocation of different foraging roles to men and women.

Men and women segregate their jobs very thoroughly – even when they share them. In medieval France, the slaughter of a pig was a task carefully divided between women and men according to custom. The woman chose the pig to be killed; the man picked the day of the killing; and so on down to sausage making (by women) and lard-salting (by men).[8] To this day women and men largely end up doing different work. Even in the Nordic countries, where nearly eighty per cent of women are in the work force, there persist clear distinctions between men's work and women's work: fewer than ten per cent of women work in occupations where the sex balance is roughly equal; half of all workers are in jobs where their own sex accounts for ninety per cent of employees.[9]

The question then arises: when did male hunting change from being just a seduction device to being part of a deal with one wife? In effect, there came a moment when men gathered meat not just to seduce more women but to feed their own children. One school of thought is that the sexual division of labour was a critical feature of our early evolution as a species. Without it we could not have survived in the dry grasslands that were our natural habitat as a species. We were too bad at hunting to make a living by it alone, and the food to be got from gathering was too unreliable and protein-deficient for our large bodies and omnivore guts. But put the two together and you have a viable lifestyle. Add cooking, which is a form of predigestion enabling us to eat tough vegetables that would normally reward only stronger stomachs than ours, and you have a viable niche for a large and social savanna ape.

Australia, New Guinea, southern Africa and parts of Latin America still contain hundreds of tribes that subsist on what they can catch and find. Most have now been bothered by anthropologists and a generalization holds true of them all: men hunt and women gather. The proportions can vary, of course. Eskimos eat a diet of pure meat, largely supplied by men; the !Kung of South Africa eat a diet of up to eighty per cent vegetable matter, supplied by women. But with just one partial exception, nearly all meat is caught by men; and nearly all vegetable food is gathered by women. The exception is the Agta people in Luzon, in the Philippines. Agta women are enthusiastic and efficient hunters, though still less so than men. Yet the Agta are not true hunter-gatherers; they trade meat for farmed food with other people.

So widespread is this distinction that where women do regularly procure meat among hunter-gatherers, it is nearly always small mammals, shellfish, fish, reptiles or grubs – prey that is caught by digging or gathering rather than ambush or chase. There is often a taboo on women handling or making weapons or hunting equipment or even accompanying hunts, but it seems most unlikely that the taboo caused the division of labour, rather than the other way round. Nor is it convincing to argue that the sexual division of labour is merely a reflection of biology, with women confined by their pregnancies and dependent children to safer, slower and less distant activities. This is much too negative a way of looking at the issue. Rather, the invention of a division of labour was an economic advance because it enabled human beings to exploit two different specializations, the results being greater than the sum of their parts. It is exactly the same argument as the division of labour between cells in a body.[10]

There is, however, a different school of thought: that until the last 100,000 years there was no sexual division of labour. Men and women were both self-sufficient foragers. Men were probably far more carnivorous than women, but there was no institution of marriage, nor any larger, band-wide pattern of food sharing to exploit the advantages of a division of labour – to make the gains from trade. We may never know how recent the switch was, but that the appearance of marriage and nuclear families within the tribe was symbiotic with food sharing is highly plausible.[11]

Food sharing is what makes it possible for men to hunt. Without sharing, human beings would not hunt, because they would not obtain enough calories that way. In many tropical hunter-gatherer societies, the caloric rewards of gathering outweigh those of hunting. Yet hunting and meat hold on to men's hearts in a way that belies their importance as a source of calories. Hunting for meat is seen as the man's principal task even in societies where he also spends much time gathering. In one part of Uganda a skinny chicken is worth the same as four days' worth of gathered plantains.[12]

The Huia bird of New Zealand, it was said in the last century, died of grief if you killed its mate. We will never know if this was more than a fable because the whole species went extinct in 1907, but we do know that Huias shared with us the sexual division of labour. Male Huias had short, strong beaks for breaking rotten wood in search of insects; females had curved, slender beaks for exploring crevices. Between them they chiselled and probed their way to food in a unique cooperative partnership between the sexes. For them, as for us, the division of labour depended upon marriage.

And like Huias, we may have developed different bodies and minds to suit the different lifestyles of our two sexes. Hunting and gathering may have left their respective marks on us. Men are innately better at throwing things than women; they are on average more carnivorous (women are roughly twice as likely to be vegetarians as men of the same age group, a discrepancy that is, if anything, increasing); and they generally prefer large meals to frequent snacks. These may be features of a hunting lifestyle. Likewise, men prove consistently better at map reading, learning their way through mazes or mentally rotating objects to see how they fit together. These are exactly the skills a hunter would need to make and throw projectiles at animals and then find his way home. Hunting itself is an overwhelmingly male occupation even in Western societies. Women are more verbal, observant, meticulous and industrious, skills that suit gathering.

There is, incidentally, abundant material for those who like stereotypes here, but none of it says anything about the woman's place being in the home. After all, the argument goes that men and women both went out to work in the Pleistocene, one to hunt, the other to gather. Neither activity was remotely like trooping off to an office and answering telephones all day. Both sexes are equally unsuited to that.

Egalitarian apes

Yet, intriguing as this tale of sexual cooperation is, it is not the most far-reaching consequence of the invention of food sharing. Giving a dead rabbit to your wife, or blackberries to your husband, is not a very surprising thing to do. The family is a cooperative unit held together in our species, as in so many others, by genetic nepotism. The couple has a shared genetic interest in its children; like ants and bees this gives them every reason to cooperate. A division of labour over food is merely another way of expressing this cooperation.

But people do not share food with just their spouses and children. They invite unrelated friends to dinner. They lunch with business partners or even rivals. They share food, if not universally, then certainly with much more largesse than they share sexual favours. If food sharing was crucial to the development of the close pair bond between husband and wife, might it also have played a role in the development of human society generally? Is virtue a shared box of chocolates?

The sharing of food is not confined to human beings. Prides of lions and packs of wolves eat their kills in communal togetherness if not entirely in harmony. But in such cases a strict rule of hierarchy still obtains. Senior wolves in the pack do not tolerate juniors taking meat from them; they merely allow them access to parts of the carcass that they are not themselves eating. Food sharing among people is something different; it is the donation of choice morsels, often fairly equitably, to others. Indeed, it is absurd to think of a dominance hierarchy at a human feast at all. Certainly, the medieval lord got the better joints brought to him than the vassals at the foot of the table. But the remarkable thing about human feasting is how egalitarian it is. The whole point of a meal is that everybody shares it equally.

Moreover, in the long story of human evolution, the invention of pair bonds between male and female is a relatively recent phenomenon, a peculiarity of our species that we share with few of our close relatives. The bonds between males in our society are much more ancient, because it is characteristic of apes, and of chimpanzees and human beings in particular, that males live in groups with their relatives, but females leave the group of their birth. In this we are wholly different from monkeys, which practise the opposite habit: females live with their kin, while males move from their native troop. Therefore, arguably, the tendency for men to feast together may go back farther than the tendency for men to share food with their spouses; it may be a legacy of sharing between related male apes.

This egalitarianism around food is something we certainly share with chimpanzees. Chimps suspend their pecking orders during a shared feast. Young and junior individuals beg for food from senior ones and usually are given them. True, the alpha male may occasionally monopolize the corpse of a monkey that has been killed, but this is by no means normal. Senior monkeys never allow subordinates to take food that is already in their possession, unless those subordinates are their close relatives. Senior chimps regularly do so, and, what is more, the juniors request food – something a junior monkey has never been seen to do, except from its mother. Chimpanzees use a whole range of gestures that are specifically related to food. They hoot to announce the discovery of a rich pile of fruit, as if calling their friends to the feast, and they beg with eloquent gestures that their friends share it with them. This is not to say they share all food always – far from it. But they do sometimes.

Frans de Waal took advantage of this among the chimpanzees of the Yerkes Primate Center in Atlanta. He delivered into the animals' enclosure bundles of fresh leafy branches from sweetgum, tulip tree, beech and blackberry, each bundle tightly bound with honeysuckle vines, and made sure that these sometimes fell into the hands of subordinate individuals. He then watched carefully what befell the contents of the bundles. He chose foliage because high-energy food, such as bananas, occasionally provoked violence among the apes, whereas foliage, though popular as food, was not quite so intensely desirable, and was often therefore shared. Whoever had a bundle would allow others to take branches from it or would give them away itself.

The first response to the appearance of the bundles was the familiar increase in general celebration that chimps display in the wild when they find a good source of food. They kissed, embraced and called. (Bonobos, or pygmy chimpanzees, the closely related species from central Africa, incidentally go one stage further when they find a rich fruiting tree: they all have sex with each other to celebrate.) The next thing that happened was an increase in 'status confirmation displays'. In other words, just before the dominance hierarchy in the group is to be suspended it is confirmed and reasserted. There is also during the feast an increase in aggression and general bickering.

None the less, the sharing is remarkably egalitarian. Dominant individuals are more likely to give than to receive. Rank matters less than reciprocity. If A often gives foliage to B, then B will often give to A. There is a pattern of turn-taking: A is more likely to give food to B if B has groomed A recently, but not if A has done the grooming favour. A chimp will punish another that has been stingy by attacking it.

To de Waal all this implies that chimpanzees 'possess a concept of trade'. They are not sharing food with each other just because they could not really prevent the others getting some of it – else, why would dominants give to subordinates? – but sharing in order to curry favour, receive reciprocal benefits in future and generally defend their reputations for virtue. They sound like sensible game theorists. 'Sharing among chimpanzees,' writes de Waal, 'is embedded in a multi-faceted matrix of relationships, social pressure, delayed rewards, and mutual obligations.'

But chimps almost never voluntarily hand over food. Sharing occurs in response to a request. So while de Waal believes they have travelled some way from the selfishness of monkeys and so garnered the benefits of reciprocal altruism, they have not, he thinks, 'crossed the evolutionary Rubicon' of reciprocity that human beings have.[13]

Spreading the risk

Above Kim Hill's desk at the University of New Mexico hangs a huge photograph of an Ache man in Paraguay with the hacked-off head of a large tapir over his shoulder. Blood is pouring down to the man's bare buttocks and thence trickling down the backs of his legs. Hill and three colleagues have revolutionized the study of human food sharing; in doing so they are unearthing the roots of economics.

It all began at Columbia University in New York in 1980. Although trained as a biochemist, Hill had worked during the previous two summers in Paraguay for the United States Peace Corps and had now come to the university to do a graduate degree in anthropology. Hill argued with a fellow student, Hillard Kaplan, about the roots of human society, trying to persuade him that anthropology was going down a blind alley because of its obsession with societies. Societies, said Hill, do not have needs, individuals do; and societies are the sums of individuals, not entities in themselves. Therefore only by understanding what made sense for individuals would anthropology make progress.

Food sharing, for instance, was at the time explained by anthropologists mainly in terms of the good of the society or the group, rather than the individual. They argued that people in tribal societies shared food with each other as a deliberately egalitarian ploy: it helped to eliminate status differentials. That in turn helped the society remain in ecological balance with its environment by discouraging people from striving for too much success in food gathering. There would be little point in gathering more than a certain amount, for they would only have to give it away. Like most social scientists, anthropologists did not feel the economist's obsessive need to explain away benevolence.

Impatient with such reasoning, Hill convinced Kaplan, and persuaded him to accompany him back to Paraguay in 1981 to begin a study of the Ache. Kaplan admits that he knew little of the theory behind anthropology, and in particular he was not yet under the influence of the great Harvard study of the hunter-gatherers of the Kalahari, the !Kung. This was crucial, for Hill's and Kaplan's ideas were to set the study of food sharing in a different direction. Two talented women now entered the picture: Magdalena Hurtado, a Venezuelan also studying at Columbia University, and Kristen Hawkes, who had first met the Ache in the 1970s. Hawkes was trained in economics and anthropology, but she was determined to use some of the ideas then coming out of biology to understand how human beings make decisions. Fifteen years and many studies later, Hawkes disagrees with Hill, Kaplan and Hurtado fervently but amicably about why hunters share their food. The next chapter is an account of their argument.

The Ache are a small tribe of nomads who depended until recently almost entirely on hunting and gathering in the rain forest. Only in the 1970s did they come into regular contact with modern society as the government of Paraguay settled them in mission camps; but in the 1980s they still spent a quarter of their time on long trips through the forest gathering and hunting. They all set out in the morning in single file until, after about half an hour, the men fan out into the forest while the women and children continue slowly along the agreed route to the evening rendezvous. The men are looking for honey or game. If they find honey, they call the women to the site and leave them hacking it out of the tree cavity where they found it. Early in the afternoon the women make camp, and collect food from the nearby forest – usually either insect grubs or the starchy pith of a palm tree. The men then arrive bringing small game such as monkeys, armadillos and pacas, and occasionally larger beasts such as peccaries or deer. Most such animals have been caught in a cooperative manner, one man calling to another for assistance when he has sighted the quarry.

Nobody is suggesting that this is how all of our ancestors lived. One of the features of human beings is their ability to adapt to local conditions, and the Paraguayan rain forest is as different from the African savanna, or the Australian desert as it was from the steppes of ice-age Europe. But what interested Hill, Kaplan, Hurtado and Hawkes was how these non-agricultural people solve what is a universal problem: the cooperative sharing and division of the spoils of hunting. They did not claim that the solution would prove universal; only that it would explain the Ache.

The Ache are astonishingly egalitarian. Although back at the settlement they tend to share only with other members of their family, while on overnight hunting trips in the forest, they share freely and widely among non-relatives in the band. The man who hands out the food is not usually the one who killed the animal. The man who returned empty-handed from the forest is not left out of the feast. Three-quarters of what anyone eats was usually acquired by somebody outside the immediate family. However, this generosity is largely confined to meat. Plant food and insect grubs, by contrast, are not usually shared outside the nuclear family.

A similar pattern of generosity obtains among the Yora of Peru. On a fishing trip, everybody shares; back at the camp, food is freely shared only in the family, and at all times meat is more widely shared than vegetables. Thus, while fish, monkeys, alligators and turtles are shared, plantains are hidden in the forest until they ripen to prevent neighbours stealing them.[14]

Why the difference? What is so special about meat that it must be shared more than fruit?

Kaplan thinks there are two plausible explanations. The first is that meat is cooperatively acquired. Monkeys, deer and peccaries are caught by the Ache after several hunters join in the pursuit, but even armadillos are usually caught when one man helps another to dig the quarry out of its burrow. Likewise, among the Yora of Peru, the man poling the canoe along the river is essential to the fishing, but does not actually catch anything himself – so it is only sensible that the fish are shared with him. Just like lions, wolves, wild dogs or hyenas, men are cooperative hunters who depend on each other for success and simply cannot afford not to share the results. They are more flexible than lions because of their specialist division of labour. One might be good at spearing fish or digging out armadillos so he specializes and his colleagues fill the other roles. As usual we find that what makes human beings unique is the division of labour.

There is another explanation of why meat is shared more than vegetables. Meat represents luck. The reason a man comes into camp with two armadillos, or a large peccary, is that he was fortunate. He might also have been skilful, but even the most skilful hunter needs luck. Among the Ache, on any given day of hunting, forty per cent of the men fail to kill anything at all. A woman who brings back little palm pith from the forest, on the other hand, was not unlucky; she was probably idle. There simply is not the same dependence upon chance for the gatherer as there is for the hunter. Therefore, sharing spreads the risk as well as the reward of hunting. If a man were to rely on his own resources he would often go hungry and occasionally have more than he could eat. But if he were to share his meat and in return expect others to share with him, he could be fairly sure of getting at least some meat every day. The sharing of meat therefore represents a sort of reciprocity in which one man trades in his current good luck for an insurance against his future bad luck – in exactly the way that vampire bats do when granting their neighbours a share of their blood meals, or that bond traders do when they swap fixed for variable interest rates.

This phenomenon is exacerbated in the tropics, where meat storage is simply not a viable option, because of the speed with which meat goes rotten. Sharing is a very effective way of reducing risk without reducing overall supply. According to one calculation, six hunters who pool their game will reduce the variability in their food supply by a massive eighty per cent compared with six hunters who do not pool their game. This is known as the risk-reduction hypothesis for food sharing.[15]

But there is a problem. What is to stop the idle from exploiting the generosity of the diligent? If you can rely on getting some meat from whoever caught it, you might as well sit by the trail and pick your nose until the hunter gets back from the forest clutching a dead monkey. The more people share their food, the more opportunity there is for the egoist to exploit the gullible and be a 'free-rider'. We are back, in a sense, with the prisoner's dilemma, but this time on a plural scale. To use a well-worn example: who will pay for a lighthouse when the light is free for all to use?

CHAPTER SIX

Public Goods and Private Gifts

In which no man can eat a whole mammoth

There is no duty more indispensable than that of returning a kindness. All men distrust one forgetful of a benefit.

Cicero

Most of the land surface of this planet is naturally desert or forest. Were it not for the actions of man, rain forests would choke the tropics, deciduous woodlands would blanket the temperate latitudes, pines would cover the mountains, spruce and fir would lie like felt across the north of Asia and North America. Only in a few places – the savannas of Africa, the pampas of South America, the steppes of central Asia and the prairies of North America – does grass dominate the ecosystem.

Yet we human beings are a grassland species. We evolved on the African savanna and we still try to recreate it wherever we go: parks, lawns, gardens and farms are all more or less managed for the benefit of grass. Indeed, as Lew Kowarski first suggested, you could plausibly argue that grass is the master of the planet, because it has employed us as its slave. We plant it, in the form of wheat and rice, where once forests stood. We tend it and loyally fight its enemies.[1]

Grass is a relative newcomer to the planet, first appearing about 25 million years ago, roughly the same time as monkeys became distinct from apes. Grass grows from the base of the plant, not the tip, so it is not easily killed by grazing. Therefore, it does not divert its precious energy into defending itself with toxic chemicals or spines; it just resigns itself to frequent setbacks at the teeth of hungry mouths. No matter; the more it is grazed the more nutrients are recycled in the dung of the grazers and the faster the grass can regrow after winter or drought.

Therefore, wherever grass grows, large animals abound. The Serengeti teems with wildebeest, zebra and gazelles, busy mouths turning grass into meat. The prairies once swarmed with herds of buffalo. By contrast, in the rain forest or the spruce forests of the north or the oak woods of temperate latitudes, large animals are few and far between; there is less for them to eat. On the grasslands, however, killing big game becomes a viable way of life for many carnivores: wolves, wild dogs, lions, cheetahs and hyenas, to name just the ones that have survived to the modern day. Notice that all of those predators – with the partial exception of the cheetah – are highly social. To bring down large game on grassland plains both requires cooperation and, because the prize is large enough to feed many mouths, allows cooperation.

This was the world in which human beings evolved. With our bipedal gait, our shade-maximizing posture, our sweat glands and bare skin, our special blood vessels for cooling the brain and our free hands for carrying things, we are superbly adapted to living in the open, sun-scorched grass plains of Africa. We are a savanna animal. We are as good at running long distances as our cousins the chimpanzees are at climbing trees. And from the earliest records, we were also hunters of large game. Stone tools and fossils of the bones they were used to dismember lie together at sites of ancient butchery deposited 1.4 million years ago or more. Careful experiment has proved to the satisfaction of most that the association was not coincidental; our ancestors ate large animals. We were also, like hyenas and lions, highly social.[2]

At the height of the ice ages, between 200,000 and 10,000 years ago, grasslands covered much of the land area of the earth. As more and more water became locked up in the ice caps and glaciers, the sea level fell, the climate dried out and rain forests shrank to small patches to be replaced by savanna. In the north, the droughts punished the trees (which are ninety per cent above ground) but benefited grass (which keeps ninety per cent of itself below ground). There were hardly any spruce forests or mossy tundras as there are today, just vast, open plains of rich grassland. These northern grasslands are known collectively as the 'mammoth steppe'. Stretching from the Pyrenees across Europe and Asia and over the great plains of Beringia (the land now mostly submerged under the Bering Strait) to the Yukon in Canada, the mammoth steppe was the greatest habitat on the planet.

We African grasslanders followed our masters, the grass, into the great mammoth steppes and took to a life principally dependent on hunting. The mammoth steppe was a grassland characterized by, and perhaps even created by, mammoths. The hairy elephants shared the habitat with woolly rhinos, wild horses and giant bison as well as smaller game, including large deer (giant wapiti), reindeer and saiga antelope. Lions were common, as were wolves, predatory short-faced bears and sabre-toothed cats. It was like a cold Serengeti.

Out on the mammoth steppe, we African grassmen felt at home (if a little chilly). We killed large animals as we had done at home. Indeed, we seem almost to have specialized in killing the largest animals of all. The Clovis people, who were among the very first into North America, were especially fond of mammoth meat. Virtually every Clovis site known contains mammoth bones. In what is now eastern Europe, 29,000 years ago, the Gravettian people made almost everything they left behind from mammoth tusks and bones: spades, spears, the walls of their houses. Our attention was too much for the mammoths. There is little doubt that the great grass-eating elephants were eventually exterminated by human hunting. This in turn hastened the disappearance of the steppe itself. Without heavy grazing and manuring, the grasslands' fertility dropped and grass began to give way to mosses and trees. These in turn insulated the ground against deep summer thaws, further depressing fertility. A vicious circle began, and rich steppes became austere tundras and taigas.[3]

Even if you have never tried killing an elephant with a spear (I have not), you will appreciate the skill of these people. We may never know their techniques for sure; they may have ambushed their prey at water holes (many carcasses are in wet areas); they may have driven them over cliffs; they may have lured them into swamps. They may even have semi-domesticated them, though it seems unlikely. But whatever they did, they did not do it alone. Cooperation was surely the key to their success. Sharing the meat was not just encouraged – it was impossible to prevent. A dead mammoth was essentially public property.

However, this brings us back to a familiar problem. Why bother to join the hunt? Why not simply turn up nonchalantly when the carcass is being divided and help yourself to a share? After all, mammoth hunting must have been dangerous in the extreme. No individual had much incentive to close with the beast and risk his life, when he could be sure of getting a share of somebody else's carcass. He would be risking his life for the common good. How the early hunters of the pre-modern era solved this problem, we may never know. I suspect that they did not, that mammoths went largely undisturbed by the Neanderthal men that inhabited Eurasia during much of the ice age. It was no accident, I believe, that most obsessive mammoth hunters date from 30,000 years ago or less. For something vital occurred about 50,000 years ago, probably somewhere in north Africa.

This was the invention of the dart thrower, the first projectile weapon and the distant ancestor of the bow and arrow. The dart thrower stores energy like a spring, imparting extra momentum to a small spear, giving it far more momentum than a large spear thrown by hand. It was the first weapon that could be launched from a safe distance. Suddenly, for the first time, a group of men could surround a mammoth and trust each other not to hang back; all could fire their weapons in relative impunity. The free-rider problem shrank. Dangerous big game became a target.[4]

Big game hunting probably began in earnest with the invention of the dart thrower. It had profound social implications. A big animal like a mammoth is large enough to share with a large group. It is so big that sharing becomes mandatory. A carcass is in effect no longer the private property of the person who killed it, but is public property, the shared possession of the group. Big game hunting not only allows sharing, it enforces it. The risk of refusing a hungry man a share of your mammoth is too great when the hungry man is armed with a dart thrower. So big game hunting introduced humankind to public goods for the first time.

Tolerated theft

A semantic digression is necessary at this point. I have used the word reciprocity as if its meaning were transparently clear. But it is actually a rather slippery word. In the form of Tit-for-tat, it means the swapping of similar favours at different times. But anthropologists have been using the word reciprocity in a subtly different sense for decades. To them it means the sharing of different favours at the same time. When a vampire bat shares a blood meal with another, it expects a blood meal in return at a later date. When a shopkeeper gives a bag of sugar to a customer, he expects money in return at the same time.

This may seem a pedantic distinction, but I believe it is vital to what follows in this and later chapters. Only under fairly unusual circumstances are two people in a position to make use of the first kind of reciprocity. Chance must supply one with a temporary benefit that the other needs; chance must then reverse the debt. And all the while each must remember the exchange. It is far easier to imagine the second kind of reciprocity, in which one person who finds himself in temporary command of a surplus can swap it for some other currency with a second person. The debt is immediately discharged, and opportunities for cheating are fewer. Imagine if to buy sugar from a shop you had to pay in sugar at a later date.

With this distinction in mind, I now turn to the argument between Kristen Hawkes and Kim Hill about why hunter-gatherers share meat with each other. Hill maintains that it is all a matter of reciprocity, in which the sharer receives some direct payment for his generosity. Hawkes considers that the reward is far more intangible, and that the sharer seeks general social recognition for his public-spiritedness in the same way that a Victorian philanthropist looked for his knighthood. The two positions are not that far apart, but it is worth exploring the debate in some detail for the light it sheds on the meaning of the word reciprocity.

The argument centres on a people called the Hadza, who live in wooded savanna country south and east of Lake Eyasi in Tanzania.

Like the Ache, the Hadza now live on the fringes of the agricultural world, occasionally taking part in it as labourers for others, but still preferring to pursue their old tradition of hunting game and foraging for roots, berries and honey. Despite the blandishments of government and missionaries, many are still (or have again become) full-time hunter-gatherers. The women forage in much the same way as Ache or !Kung women, seeking out tubers, fruit and honey – usually from wild bee colonies that men have located during hunting trips. But the Hadza men, unlike the Ache or the !Kung, set out to kill really big animals with their bows and arrows – usually antelopes, but occasionally up to the size of giraffes. A giraffe carcass contains a vast amount of meat, far more than a single man can possibly consume or store in the African sun. So the lucky hunter has little option but to give the meat away to his friends, who therefore stand to gain from his selfless act of going out to hunt. The question he must ask himself is why he bothered. It probably took him several months of hunting before he killed a giraffe, whereas he could have caught a guinea fowl several times a week if he had set snares for them. He could then have kept the fowl for his family and need not have shared them with his neighbours.[5]

Kristen Hawkes asked Hadza men to try catching small game, such as guinea fowl, using snares and traps. They got less meat altogether, but got at least something on many more days. On average, when hunting big game, they came back empty-handed on ninety-seven days out of 100. So Hawkes concluded that a sensible Hadza man, interested only in the welfare of his children, should take to a life of snaring so that he could be sure of putting meat on his family's plate almost every day. That surely would please them more than half a ton of steak every six months. But this is not what they do. Hawkes seeks to understand why not.

Moreover, since anybody who kills a giraffe is virtually bound to share it freely, the sensible man merely waits at home until he hears the good news that another, more public-spirited person has brought home the bacon. The larger the carcass, the less of it the hunter will keep, yet the Hadza persist in chasing big animals that they will mostly give away. So why are they such generous sharers?

Hawkes believes food sharing is little more than 'tolerated theft', a term coined by her colleague Nick Blurton-Jones. Once the man who killed the giraffe has hacked off as much meat as he can carry, he has little incentive to prevent others helping themselves; to defend the carcass against them would be spiteful and inconvenient. This idea originated with Glyn Isaac, an anthropologist who suggested in the 1960s, shortly before his untimely death, that food sharing occupied a central place in human evolution but that it evolved out of tolerated scrounging of the kind seen in animals. Lions, for example, are plainly tolerated thieves: at a lion feast, God helps those who help themselves. Chimpanzees are a little more genteel, but they still have to beg for food, whereas human beings can expect to be offered it. Developing this idea further after studying the Hadza, Nick Blurton-Jones later came to argue that tolerated theft was not just a stage ancient proto-humans had passed through, it was a still-valid description of why hunters share meat with their comrades. Blurton-Jones noticed an edge of hostility in the process of sharing food among the Hadza.[6]

The logical way to view a large carcass killed by a Hadza hunter, therefore, is as the oldest example in the world of a 'public good': something provided for the benefit of the community. A public good poses what is termed the collective-action problem, which is nothing less than our old friend the prisoner's dilemma writ large. A lighthouse is the classic example of a public good. It is erected at some expense, but its light can be used freely by anybody to guide his ship to port, even if he refused to subscribe to the building of the lighthouse. Therefore it is in everybody's interests to let everybody else pay for the lighthouse, so lighthouses do not get built – or rather, they do, but it is not immediately clear why. A dead giraffe, Hawkes reasoned, is a bit like a lighthouse: it takes somebody to catch it, but when it is caught the meat is simply there to be shared by even the laziest member of the camp before it goes rotten.

So why, asked Hawkes, do hunter-gatherers work at all? She turned to the work of an American economist of the 1960s, Mancur Olson. Olson argued that the problem of providing public goods can easily be solved if there are sufficient social incentives. The successful merchant, anxious to enhance his standing and reputation in the town and prepared to spend a little money on it, announces that he will pay for the lighthouse. Precisely because this is a munificent act that will benefit others, it grants him kudos.

Likewise, the Hadza men who are good at hunting enjoy considerable social rewards. Their success is envied by other men and, perhaps more important, admired by the women. Good hunters, to put it bluntly, have more extramarital affairs. This is not confined to the Hadza. It applies to the Ache, the Yanomamo and other South American tribes; it is probably universal and it is no secret.

This may explain why men are so obsessed with killing big, shareable items. It is a noticeable feature of male human beings that wherever they live they seem to seek out the kinds of food that must be shared widely even at the expense of ignoring some rather more profitable, smaller prey. Look at it from the hunter's point of view. If he kills a guinea fowl, his wife and children eat it; if he kills a small antelope, there might be a bit left over for his creditors among the other hunters. But if he kills a giraffe, there is so much meat that nobody will notice him slipping a choice cut to the nubile wife of a neighbour.

Of course, this merely shifts the puzzle to the women. The male incentive for chasing giraffes when they could be gathering guinea fowl for their families is suddenly clear: it leads to sex. They are more interested in supplying their mistresses than their children. But why does it lead to sex? Why do women reward hunters with affairs? Here is where Hawkes disagrees most plainly with Kaplan and Hill. Hawkes says the attraction is an intangible one; the mere smell of success, which she calls 'social attention', is attractive to the women. They get nothing from the deal save a nudge upwards in status. Hill and Kaplan say otherwise. They argue that there are very tangible benefits for the women: choice cuts of meat. Not all parts of a giraffe are equally tasty, and the hunter who killed it can easily monopolize the best bits and use them directly to bribe women with whom he wishes to have an affair. The mystery of why he does not bother with guinea fowl is therefore easily solved, and food sharing, far from being done under duress, is a directly reciprocal act just as it is in chimps and in the Ache. We are right back with the male chimpanzees of Gombe (which is not far from Hadza territory), setting out to catch a monkey to feed to a sexually receptive female. The reciprocity comes in a different currency – sex.

In any case, Hill and Kaplan challenge Hawkes's premise that the men would be better off catching guinea fowl. So long as the meat from large game is shared, the Hadza men actually eat considerably more calories if they chase big game than if they chase small game. The extra size of the carcasses more than compensates for the infrequency with which they are caught. In the case of the Ache, Hill and Kaplan calculate that hunting peccaries produces about 65,000 calories per hour of work, whereas searching for insect grubs is much less rewarding, producing 2,000 calories per hour. True, you have to share a peccary with the rest of the band – and on average get to keep only about ten per cent of the meat – whereas you will have to share only sixty per cent or so of the grubs you find. But ten per cent of 65,000 calories is still more than forty per cent of 2,000. So it still pays Ache men to hunt pigs rather than gather grubs.

Hill and Kaplan argue that 'nothing in Hawkes's review of the data suggests that hunters do not simply exchange meat for other goods and services. This is crucial, because if such trade is common, large game does not constitute a public good and no collective-action problem exists.'[7] In most hunter-gatherers there is a pronounced bias in food sharing; the nuclear family of the hunter takes a disproportionate share, especially of small carcasses, suggesting that – contra the tolerated-theft hypothesis – the hunter does retain some control over the destination of the meat. In the Gunwinggu of Arnhem Land in northern Australia the successful hunters do end up with more meat for their families than the others, and they go to great lengths to favour kin over non-kin. In the Ache, food is sometimes kept for those who were absent from the sharing. And oddest of all, the man who killed an animal usually eats less than his share of it. These features do not suggest the contest over meat that tolerated theft implies.

It is a question of who has the power: the haves or the have-nots. If sharing is tolerated theft, the have-nots are powerful; if it is reciprocity, the haves are in control. Even if the Hadza hunter knows he will eventually lose the giraffe to tolerated theft, he can still influence the sharing; his aim is to turn the sudden surplus of giraffe meat in his possession into some less perishable currency. So he shares it with his spouse and kin; with potential mates; and with his friends from whom he has had, or expects to have, a reciprocal favour. This evens out his supply of meat by giving him to expect a share of others' carcasses in the future. And it buys him prestige.

Hawkes replies to these charges with some telling darts of her own. She says there simply is no evidence for the strict reciprocity of the Hill–Kaplan world. Bad hunters and free-riders are not punished. Yet there are consistently idle or incompetent individuals. They lose social attention, yes, but they do not lose meat. Why are the other men feeding them?

The social market

And so the debate continues. It probably reflects some genuine cultural differences between the Ache and the Hadza, or even the different genders of Hawkes and Hill. Yet, at the risk of annoying both sides, I think they are saying much the same thing. Hawkes is saying that the payback to a good hunter is not meat but prestige; Hill and Kaplan are saying that he hunts because there is a payback. The argument is an echo of a much older debate in anthropology between the 'substantivists' and the 'formalists'. Like all disagreements in academia, it raged so fiercely at its height in the 1960s and 1970s largely because the stakes were so small – there was only the subtlest of differences between the positions of the two schools. The formalists argued, like Hill and Kaplan, that the insights of economics are applicable to tribal societies, and people's decisions in those societies can be analysed just like those of people in Western market-based countries. Thus, for a formalist, the origin of the market, with all its capacity to exchange goods of different kinds, exploit the division of labour and provide a hedge against dependence on one good, may lie in the reciprocal food-sharing arrangements of a hunter-gatherer band.[8]

The substantivists, however, say that economics cannot apply to primitive societies because the people in those societies are not in a market at all. They are not free agents, deciding their own self-interest in the passionless world of a shopping mall. They are embedded in a tangle of social obligations, kin networks and power relations. The reason a person shares food with another may be because of a calculated reciprocal hedge, but it might also be because he is bound by custom to do so, or intimidated into it by his fear of the recipient's power.

Hawkes, in the substantivist tradition, rebels against the naked economics of reciprocal sharing. As I say, this is surely hair-splitting; modern economics also tries to broaden its attention beyond the perfect market and take into account the 'irrational' reasons people have for their decisions. And even if Hawkes is right that Hadza men hunt for the prestige rather than the return favour, you can still take a ruthlessly economic view of their motives: they are converting giraffe meat into a durable and valuable commodity – prestige – that will be cashed in for a different currency of advantage at a later stage. For this reason, Richard Alexander calls the trading of concrete for abstract benefits 'indirect reciprocity'.[9]

Indeed, to take this argument a little further, I do not believe it is too far-fetched to see in the actions of hunter-gatherers distant echoes of the origins of modern markets in financial derivatives. When a Hadza man shares meat with the expectation of some future return, he is in effect buying a derivative instrument with which to hedge his risk. According to Hill and Kaplan, he is entering into a contract to swap the variable return rate on his hunting effort for a more nearly fixed return rate achieved by his whole group. He is just like a farmer who contracts to receive a fixed income for his wheat in six months' time by selling a forward contract or buying some futures. Or like a banker who has lent a large loan at a variable rate of interest, and decides to hedge his position by signing a contract for a swap (or perhaps even a swaption – an option to swap) with another bank: he agrees to pay a series of variable payments, linked to short-term interest rates, in exchange for receiving a series of fixed payments. In doing so he seeks out a counterparty who wants the opposite.

According to Hawkes, the hunter is reducing his exposure to one currency (meat) by buying another (prestige), in just the same way that a company that can raise a loan cheaply in dollars might swap it for one in Deutschmarks to hedge its exposure to exchange rates. The analogies are far from exact, but the principles are precisely the same: one person wishes to reduce his risk by trading with another, or with others. Those tempted to scoff at hunter-gatherers for being far too unsophisticated for this sort of thing would be wrong. Their brains are the same as ours, and their instincts for good deals are as closely honed within their own cultural environments as those of any broker on the Chicago Mercantile Exchange. And by seeing it in this light, an important insight emerges. The defence that derivatives traders give for their trade is that they are in the business of reducing risk by matching together individuals who have different exposures. They argue that a futures market or a swaps market benefits everybody. It is not a zero-sum game. If they are not able to swap risks, businesses are exposed to more risk, for which they have to pay. Exactly the same argument applies to the origin of hunting and food sharing in human beings. Hunting is risky; sharing reduces that risk. Everybody benefits.[10]

If the Hadza seem too remote, consider a similar problem closer to home: windfalls of good luck. There are many examples of people who have experienced sudden good fortune and have been deeply resented in their communities for not sharing it with others. One San woman who was well paid for her part in a film called *The Gods Must Be Crazy* spent it all on things for herself, and so provoked a fight.[11]

Likewise, Marshall Sahlins argued that the reason hunter-gatherers are so generally idle – they 'work' far fewer hours than farming people – and so free of possessions and wealth, is because in their egalitarian societies to accumulate too much is to refuse to share it, so it makes better sense to want little and thereby achieve all they want. Hunter-gatherers, said Sahlins, had discovered the Zen road to affluence; they work hard enough to provide for their various ambitions and needs; then, rather than risk jealousy, they stop.[12]

On 8 August 1993, Maura Burke won £3 million in the Irish national lottery. The 450 people who lived in the tiny village of Lettermore were delighted for their fortunate neighbour and threw a spontaneous party. Mrs Burke's husband died within a month and she had no children. Expectations ran high in the village. Yet she did not share anything with the villagers, and they quickly grew resentful. 'We've not seen a penny of it,' one resident said angrily to a journalist. Mrs Burke began to receive death threats and moved to London. Her good fortune had driven her out of her community because she was unwilling to share.[13]

At first sight, Mrs Burke's punishment was very much in the tradition of Hawkes's tolerated theft. The community did not just expect her to be generous with her windfall, it punished her for not being generous. Yet there is another way to look at it: Hill's and Kaplan's way. Like a player in a prisoner's dilemma game, Mrs Burke had suddenly defected after cooperating for many years, and her partners felt inclined to punish her. Knowing the neighbours could never offer her the same generosity in the future, she had little incentive to share. But a fortunate aboriginal hunter knows it is only a matter of time before he finds himself in the position of recipient rather than donor. The long shadow of the future hangs over his decision.

Incidentally, Mrs Burke was lucky. In Eskimo societies, to hoard is taboo. Rich people who are ungenerous are sometimes killed.

Gifts as weapons

At first glance this explains why human beings are such enthusiastic collaborators. Yet it is not entirely a satisfactory explanation, for reasons outlined by a brilliant Israeli scientist with a habit of putting cats among intellectual pigeons, Amotz Zahavi. He studies Arabian babblers, which, like many medium-sized birds in warm parts of the world, live not in pairs but in larger family groups in which the 'teenagers' help the parents rear more young. Such helping at the nest has never seemed to present much of a problem for biologists to explain. After all, merely by hanging around, the teenagers increase their chance to inherit the breeding role, meanwhile bringing brothers and sisters into the world. It is a system driven by nepotism and selfishness.

But Zahavi was puzzled by the enthusiasm of the teenagers. Not only do they compete vigorously to bring food to the nest, to take on the role of sentinels watching for predators and to defend the territory against intruding neighbours, but their enthusiasm seems to be strangely unwelcome. Dominant birds actually try to prevent subordinates from helping, whereas they should, Zahavi thought, free-load upon their younger siblings' efforts.

Zahavi argues that the helpers are not pursuing nepotistic or inherited rewards at all, but are after something he calls social prestige. Vigorous and energetic helping, he says, emphasizes the commitment of the bird to the family, which in turn draws similar commitment from the other partners. This leads Zahavi to a reassessment of marriage – at least in birds. 'I suggest that, even in collaborations of two, a large part of the investment can be explained as an advertisement of the quality of the investor and of its motivation to continue collaborating, in order to decrease the partner's tendency to cheat or desert.' Zahavi's conclusion depicts generosity as a weapon.[14]

Human cultures echo this strange ambiguity. At any one time in Britain, about seven to eight per cent of the economy is devoted to producing articles that will be given away as gifts. In Japan the figure is probably higher. It is a largely recession-proof industry as proved by the eagerness with which manufacturers of refrigerators and cookers diversified in recent decades into goods such as toasters and coffee-makers, items whose sales are dominated by the wedding and Christmas markets. They explicitly did so as a hedge against recessions. But why do people give each other gifts? It is partly to be nice to them, partly also to protect their own reputations as generous people, and partly too to put the recipient under an obligation to reciprocate. Gifts can easily become bribes.

Take the habit of *kula*, as practised in the Trobriand Islands. *Kula* is the exchange of shell necklaces for armbands. The islands form a circular archipelago, and people give necklaces to those on islands clockwise from them, and armbands in exchange to those on islands anticlockwise. The two kinds of *kula* goods travel in an endless circle, utterly pointless but inexpressibly important. Why is gift giving such an obsession of man?

In the 1920s, the French ethnographer Marcel Mauss wrote his famous '*Essai sur le don*', in which he suggested that gift giving in pre-industrial societies was a way of making social contracts with strangers. In the absence of the state to secure peace, gift giving served the same purpose. In the 1960s Marshall Sahlins noticed a rather obvious feature of societies all around the world. The closer the kinship between the person giving the gift and the person receiving it, the less necessary it was that the gift be balanced by a commensurate gift in return. Within the family, said Sahlins, there was 'generalized reciprocity', by which he meant no reciprocity at all: people just gave each other gifts without keeping a count of who owed whom. Within the village or the tribe, it was necessary to be fairly exact in balancing a gift. Between tribes there was what Sahlins termed negative reciprocity, his rather confusing term for theft, or for an attempt to get something for less than what it is worth. Only with unrelated allies was true reciprocity – value for value – practised.

Of course a parent does not expect reciprocal generosity from a child, and of course a thief is not expecting to pay for his loot, but in every other case, a gift is very clearly intended to be reciprocated in rough proportion. The recipient is embarrassed not to have something to give in return, or is annoyed at the thought that you might feel a small box of chocolates to be sufficient payment for all the help they have given you in some way. Even if the two payments are in entirely different currencies, the point of giving is to exchange. About the only exception, it seems to me, is sending flowers to a friend in hospital, and even there you expect him to send you flowers when you are in hospital.

The instinct is immediately familiar. Try to imagine a world without it; a world in which people did not mind how generous you were, nor did you mind how grateful they were. From deep down inside you comes this irrepressible tendency to see the world of gift giving in terms of deals (except among relatives).

As so often, this is easier to notice in cultures other than our own. When Columbus first stepped ashore in America, he met people who were separated by many tens of thousands of years from all cultural contact with the ancestors of Europeans. These two lineages had had no opportunity to transmit practices to each other since the Mesolithic. Yet there was no difficulty in understanding that gifts were given in the expectation of being reciprocated. It was one of the things that the red and white men fell instantly to doing. The term 'Indian gift' came to mean, in colonial America, a present for which an equivalent return is expected. Gifts came with strings attached – that was the whole point of gifts. To this day, it is one of the least incomprehensible cultural universals. When one anthropologist worked with a Kenyan tribe, he was struck by how they belittled everything he gave them. 'Every gift horse was examined carefully in the mouth and found wanting,' he said. But he had no difficulty understanding why. Gifts are given with an element of calculation, and his recipients knew this as well as he did. There is no such thing as a free lunch. Even in the most sophisticated European circles, you feel the obligation that comes inseparably with a rich present from somebody.[15]

Keeping up with the generosity of the Joneses

Before you accuse me of total cynicism, note that I am not trying to take the virtue out of virtue. If you worry too much about the motives of generous people, you go round in circles. A true altruist would not give a gift, because he would realize that he was either motivated by vainglory of doing good or expecting reciprocation, in which case he was unkindly putting the recipient in his debt. A truly altruistic recipient would not insult his donor by reciprocating the gift, throwing the debt back and implying that the motive was not selfless. So the truly altruistic pair never give each other anything, and only someone devoid of motives can do good. Something must be wrong there somewhere.[16]

Paradoxes aside, suffice it to say that the human instinct to reciprocate a gift is so strong that gifts can be used as weapons. Take the practice of 'potlatch', the habit of deliberately trying to embarrass your neighbours with your generosity. Although this practice is known from various parts of the world, including New Guinea, it was most famously practised among American Indian groups in the Pacific North-west until the nineteenth century. The name comes from the Chinook language. We know the details best from one tribe, the Kwakiutl of Vancouver Island.

The Kwakiutl were consummate snobs. What mattered to them above all was status, as expressed by the noble titles they tried to accumulate. What terrified them was humiliation. Their lives were dominated by the obsessive search for status and fear of shame. Deprived of the chance to make war by the Canadian government, the principal weapon they used was generosity. They distributed their wealth to earn each step up the social ladder and lost face and status by failing to repay the generosity of others with generous interest.

So ritualized was this absurd contest that special events – potlatches – were devoted to the ostentatious display of generosity and consumption by battling rivals. They gave each other blankets, candlefish oil, berries, fish, sea-otter pelts, canoes, and, most valuable of all, 'coppers', sheets of beaten copper decorated with figures. Not content with giving away wealth, some potlatch hosts took to destroying it instead. One chief tried to put out his rival's fire with expensive blankets and canoes; the rival poured candlefish oil on the flames to keep them burning. In some feast-houses, special figures carved into the ceilings, known as vomiters, disgorged a continuous stream of precious oil into the fire. The guest had to pretend not to notice the heat from the flame, even when it was blistering his skin. Sometimes, to the great credit of the host, the house burnt down.

Urging her son on to feats of generosity, one woman invoked the memory of her father:

'He gave away or killed slaves. He gave away or burned his canoes in the fire of the feast-house. He gave away sea-otter skins to his rivals in his own tribes or to chiefs of other tribes, or he cut them to pieces. You know that what I say is true. This, my son, is the road your father laid out for you, and on which you must walk.'[17]

Absurd as this sounds, it was not without method. Clearly the most ostentatious potlatches were not ordinary events, else there would never have been any wealth to give away. They were the extreme manifestations of a system of competitive accumulation of wealth. And they were distinctly reciprocal. Each gift had to be matched with interest; each feast or destructive display surpassed by another. Some of the potlatches even consisted of ritualized auctions of valuable coppers by one chief to another. But there was always a loser. In the world of the potlatch, reciprocity was not something that benefited both sides.

What possible use could this have in a rational, economic world? The formalist reply is simple: the potlatch consists of goods that are perishable or vulnerable; the prestige that it buys is a good that is durable and portable. If a chief suddenly has a glut of food or oil, he cannot preserve it so instead he holds a feast, gives it away or even, in extreme cases, burns it. This extravagance or generosity wins him respect and prestige. This does not fully explain why durable goods, such as coppers and blankets, were consumed so conspicuously in potlatches, but even here there is a logic: if coppers can buy prestige, then trade them in for it. As Ruth Benedict put it, 'These tribes did not use wealth to get for themselves an equivalent value in economic goods, but as counters of fixed value in a game they played to win.'[18]

And yet it is stretching things to try to understand potlatches as rational strategies for reaping the benefits of reciprocity. Rather, I suspect it is a selfish and devious method for exploiting the human capacity for falling for reciprocity, a sort of parasitism of reciprocity. Potlatches were designed to exploit the fact that people instinctively could not resist the temptation to return generosity.

Let me explain. Potlatches were not uniquely peculiar to the Kwakiutl and their neighbours. Competitive gift giving was a familiar way that European monarchs ingratiated themselves with each other and with oriental dignitaries. Ambassadors lost face on behalf of their countries if the gifts they brought were not sufficiently valuable. Office mates or neighbours who have received bigger Christmas gifts than they gave know the feeling. So do businessmen arriving in Japan with the wrong kind of present. The Dauphin quite clearly insulted Henry V by sending him a coronation gift of tennis balls; and you would be insulted to be given a toothbrush for your birthday. Gifts can be weapons.

Throughout the Pacific, islanders exchanged gifts in escalating battles of showing off. In 1918, for instance, an insult passed between two Trobriand Islanders from different villages about the quality of yams grown in the second village, Wakayse. The Wakayse man returned the insult to the Kakwaku man. The chiefs of the villages supported their respective plaintiffs and the dispute grew nasty. So the Wakayse men put together an enormous crate, 14.5 cubic metres in capacity, filled it with yams and delivered it to Kakwaku. The next day it was returned filled with different yams grown in Kakwaku. It could have been filled twice over, claimed the men from Kakwaku, but that would have been insulting. Peace was restored.

Malinowski's description of a typical Trobriand yam exchange, known as *buritila'ulo*, captures the far from altruistic atmosphere that surrounds gift giving in human beings. In another example, he cited the relationship between coastal fishermen and inland yam growers. The fishermen had taken up pearl diving and found it highly profitable; they could earn enough money to buy all the yams and fish they needed. But the yam growers inland insisted on giving them yams, and the fishermen had to give up their pearl diving to catch some fish to send to the yam growers in exchange. By creating obligation, the gift is a weapon.[19]

But it is only a weapon if there is a sense of obligation in the first place. Gift giving and competitive generosity is not some human invention that shaped our natures; it is a human invention to exploit our pre-existing natures, our innate respect for generosity and disrespect for those who would not share. And why would we have such an instinct? Because to be known as intolerant of and punitive towards stinginess is an effective way to police a system of reciprocity, to extort your share of others' good fortune. So gift giving in a tribal society, where the object is to put somebody else under an obligation, is not gift giving at all; it is exploiting a reciprocal instinct.

If, as I have argued, gift giving is an expression and sometimes a parasitism of the reciprocity instinct, we should be able to find and expose that instinct by experiment, just as we can find and expose a dog's instinct to salivate when it hears a signal that food is near. Can we?

CHAPTER SEVEN

Theories of Moral Sentiments

In which emotions prevent us being rational fools

The discovery that tendencies to altruism are shaped by benefits to genes is one of the most disturbing in the history of science. When I first grasped it, I slept badly for many nights, trying to find some alternative that did not so roughly challenge my sense of good and evil. Understanding this discovery can undermine commitment to morality – it seems silly to restrain oneself if moral behavior is just another strategy for advancing the interests of one's genes. Some students, I am embarrassed to say, have left my courses with a naïve notion of the selfish-gene theory that seemed to them to justify selfish behavior, despite my best efforts to explain the naturalistic fallacy.

Randolph Nesse, 1994[1]

The isolated island of Maku in the central Pacific is inhabited by a fierce, tribal Polynesian people called the Kaluame. They hold a unique place in the history of science because of two studies that took place simultaneously of the same local chieftain, an ample man known as Big Kiku. The first study was done by an economist interested in reciprocal exchange; the second by an anthropologist out to document the innate selflessness of human beings. Both experts had noticed a peculiarity of Big Kiku, that he demanded that his followers have their faces tattooed to show their loyalty. One night, just as it grew dark, four frightened and hungry men stumbled into the camp where the two intellectuals were eating their dinner in competitive silence. They asked Big Kiku to be fed with some cassava. He told them:

'If you get a tattoo on your face, then you will be fed a cassava root in the morning.'

The two intellectuals looked up, interested. How, wondered the economist, do the four men know that Big Kiku will keep his word? He might tattoo them and then still not feed them.

I simply do not believe that Big Kiku is serious, replied the anthropologist. I think he is merely bluffing. You and I know what a charming fellow he is, and he surely would not refuse food to a man just because he did not get a tattoo!

They argued late into the night over a bottle of whisky, and the sun was already high into the sky when they rose the next morning. Recalling the four hungry refugees, they asked Big Kiku what had happened. This was his reply:

'All four left at sun-up. But since you are so clever, I will set you a test, and if you get it wrong, I will tattoo your faces myself. The first man got a tattoo, the second had nothing to eat, the third did not get a tattoo and to the fourth I gave a large cassava root. Now, you tell me which of the four you need to know more about to answer your curiosity about my behaviour: the first, the second, the third or the fourth. If you ask about one that is irrelevant to your inquiry, or fail to ask about one that is relevant, you lose, and I get to tattoo your face.' He laughed loud and long.

As you have probably realized by now, there is no such place as Maku, no such people as the Kaluame and no such philosopher king as Big Kiku. But put yourself in the position of each of the two intellectuals in turn and answer the question. It is a well-known psychological puzzle called the Wason test, usually played with four cards, and you are required to turn over the minimum number of cards to test a certain if–then rule. People are surprisingly bad at the Wason test in some circumstances – for instance, if presented with it as an abstract piece of logic – but surprisingly good at it in others. In general, the more the puzzle is presented as a social contract to be policed, the easier people find it, even if the contract is deeply foreign and the social context unfamiliar.

I have slightly embroidered a version of the Wason test told by Leda Cosmides and John Tooby, a husband and wife team of psychologist and anthropologist; they invented Big Kiku and his culture in order to present people with a wholly strange world in which they could not bring their own cultural biases to bear.

The economist's puzzle is comparatively easy. About three-quarters of seventy-five students at Stanford University got it right when asked. Remember he is interested in knowing if Big Kiku kept his word. To avoid having his face tattooed, the economist must ask Big Kiku whether he gave food to the first man (who got the tattoo) and whether the second (who went away hungry) got the tattoo. The other two are irrelevant, because Big Kiku had not broken his word if he refused food to a man who did not get a tattoo, or indeed if he fed a man who did not; he simply said if the man got a tattoo then he would be fed.

The anthropologist's problem is logically similar, but it proves to be much harder. When it is posed to Stanford students, the majority of them gets it wrong, however carefully it is worded.[2] The anthropologist is looking for evidence that Big Kiku is unconditionally generous: he sometimes lets people eat when they have not got the tattoos; he does not care about those who got the tattoo. So he is only interested in the third and fourth men: he who got no tattoo (and might have been fed anyway), and he who was fed (and might not have got a tattoo). The first two are irrelevant because Big Kiku was not generous to either.

Why is the second problem so much harder? The answer goes straight to the heart of the question posed in Chapter Six: whether humans have an instinct to reciprocate, and to see that others reciprocate. The economist is looking for cheats, who do not keep their word, a familiar and easy idea that comes naturally to all of us; the anthropologist is looking for altruists, who offer a bargain and then give away their side of it anyway. Not only is that an unusual thing to find going on, it is also something that poses no threat to your own self-interest if somebody else does it. If somebody offers to buy you lunch, you do not worry about his generosity, but his normal lack of it; you worry about whether he might intend to ask a favour of you in return.[3]

The Big Kiku case was not an isolated experiment; it was part of a long series of experiments in which psychologists gradually narrowed down the question of what makes a Wason test hard and what makes one easy, itself part of the discovery that the laws of thought and the laws of logic are very different things. Familiarity with the context and the story makes no difference, they found. Logical simplicity matters very little: some complicated Wason tests are easy to solve. The fact that the puzzle is presented as a social contract *per se* does not matter, either. What matters is whether the person being tested is asked to identify cheats in social contracts – people who take the benefits without paying the cost. People are bad at looking for altruism; better at looking for cheating. People are bad at judging tests where it is hard to guess the cost and the benefit of the various actions. They are bad at looking for rewards and losses when these are not illicit in some sense. Even when the Wason test was adapted by one student for the Achuar people of Ecuador, who are almost completely isolated from contact with the Western world, there was strong evidence that they too were far better at detecting cheats of social contracts than at other forms of reasoning.[4]

In short, the Wason test seems to tap straight into a part of the human brain that is a ruthless and devastatingly focused calculating machine. It treats every problem as a social contract arrived at between two people and looks for ways to check those who might cheat the contract. It is the exchange organ.

This seems ridiculous; how can a part of the brain instinctively 'know' social contract theory? Has Rousseau somehow infiltrated the genes? It is no more absurd than arguing that the brain knows calculus because a sportsman can catch a ball by extrapolating its trajectory, or grammar because you know how to make a past tense from a previously unknown verb, or that the eye is capable of higher physics and mathematics because it slightly adjusts the colour of an object according to the general colour of the whole scene, thus correcting for the redness of the evening light. All the exchange organ does is robotically employ specialized inference engines designed by natural selection to find violations of exchange contracts agreed between two parties. As a species, wherever we live and in whatever culture, we seem to be uniquely aware of cost-benefit analysis of exchanges. We simply do not have organs that are designed to spot other, logically comparable but socially different events, such as when people have made mistakes or broken prescriptive rules that are not social contracts. Nor are we good at spotting irrational situations that defy descriptive rules of no social significance. There are people with certain kinds of brain damage who prove to have lost almost nothing except the ability to reason about social exchange; there are, conversely, people – especially most schizophrenics – who fail most tests of intelligence except those that concern reasoning about social exchange. Imprecise as the concept seems, the human animal does appear to have an exchange organ in its brain. We shall see later that neurology already supports such an outlandish idea.[5]

We invent social exchange in even the most inappropriate situations. It dominates our relationship with the supernatural, for example. We frequently and universally anthropomorphize the natural world as a system of social exchanges. 'The gods are angry because of what we have done' we say to justify a setback in the Trojan war, a plague of locusts in ancient Egypt, a drought in the Namib desert or a piece of bad luck in modern suburbia. I frequently kick or glower at recalcitrant tools or machines, cursing the vindictiveness of inanimate objects, blatant in my anthropomorphism. If we please the gods – with sacrifices, food offerings, or prayer – we expect to be rewarded with military victory, good harvests or a ticket to heaven. Our steadfast refusal to believe in good or bad luck, but to attribute it to some punishment for a broken promise or reward for a good deed, whether we are religious or not, is idiosyncratic to say the least.[6]

We do not know for sure where the social-exchange organ is, or how it works, but we can tell it is there as surely as we can tell anything else about our brains. An astonishing hypothesis has begun to emerge in recent years along the border between psychology and economics. The human brain is not just better than that of other animals, it is different. And it is different in a fascinating way: it is equipped with special faculties to enable it to exploit reciprocity, to trade favours and to reap the benefits of social living.[7]

Revenge is irrational

Biologists discovered nepotism and reciprocity in the 1960s because they caught the self-interest virus. They suddenly started asking, about everything that had evolved, 'But what's in it for the individual?' Not the species, or the group – the individual. Such a question led them to a fascination with animal cooperation and hence to the central importance of the gene. Behaviour that is not in the interest of the individual might be in the interest of its genes. Material self-interest for genes became the watchword of biology.

But a curious thing has happened in recent years. Economists, who founded their whole discipline on the question 'What's in it for the individual?', have begun to back away. Much of the innovation in economics of recent years has been based on the alarming discovery by economists that people are motivated by something other than material self-interest. In other words, just as biology shook off its woolly collectivism and donned the hair shirt of individualism, economics has begun to go the other way: to try to explain why people do things that are against their selfish interests.

The most successful of those attempts is that by Robert Frank, an economist. His is a theory of why we have emotions, founded in a combination of the new cynical biology and the less pecuniary economics. It may seem odd that a man who has written a textbook on microeconomics should steal in where psychologists have floundered, and explain the function of emotion. But that is exactly the point he makes. Human motives are the stuff of economics, whether they are rational and material or not.

Robert Trivers, who brought gene-centred cynicism to much of biology, once wrote: 'Models that attempt to explain altruistic behavior in terms of natural selection are models designed to take the altruism out of altruism.'[8] This is an old idea for social sciences, as familiar to the Glasgow philosophers of the eighteenth century as to modern economists such as Amartya Sen: if you are nice to people because it makes you feel better, then your compassion is selfish, not selfless. Likewise, in the world of biology, an ant slaves away celibate on behalf of its sisters not out of the goodness of its little heart (an organ it does not possess in a form that we would recognize), but out of the selfishness of its genes. A vampire bat feeds its neighbour for sound, ultimately selfish reasons. Even baboons that repay social favours are being prudent rather than kind. What passes for virtue, said Michael Ghiselin, is a form of expediency. (Christians should pause before they feel superior: they teach that you should practise virtue to get to heaven – a pretty big bribe to appeal to their selfishness.)[9]

The key to understanding Robert Frank's theory of the emotions is to keep in mind this distinction between superficial irrationality and ultimate good sense. Frank began his seminal book, *Passions within Reason*, with a description of a bloody massacre by some Hatfields of some McCoys. The murderers were being irrational and self-defeating in their act of quite unnecessary revenge, which in turn led to revenge on them. Any rational person would not pursue a feud, any more than he would let guilt or shame prevent him from stealing a friend's wallet. Emotions are profoundly irrational forces, Frank argues, that cannot be explained by material self-interest. Yet they have evolved, like everything else in human nature, for a purpose.

In the same way, ants that rear their sisters rather than their daughters seem superficially irrational, or for that matter mice that rear daughters rather than looking after themselves are apparently ignoring material self-interest. Yet probe beneath the surface of the individual to its genes and all becomes clear. The ants and the mice are selflessly serving the material interests of selfish genes. In the same way, Frank argues that human beings who let emotion rather than rationality govern their lives may be making immediate sacrifices, but in the long term are making choices that benefit their well-being. Notice that I am not using the word emotion here to mean 'affect': hysterical or paranoid people may seem highly irrational, but they are in the grip of an affect, rather than a specific emotion. Moral sentiments, as Frank (and Adam Smith before him) calls the emotions, are problem-solving devices designed to make highly social creatures effective at using social relations to their genes' long-term advantage. They are a way of settling the conflict between short-term expediency and long-term prudence in favour of the latter.[10]

Commit yourself

Frank's general term for this is the commitment problem. To reap the long-term reward of cooperation may require you to forgo the short-term temptation of self-interest. Even if you know that, and are determined to reap the long-term reward, how do you convince other people you are committed to such a course? The economist Thomas Schelling has dramatized the commitment problem in a story known as the kidnapper's dilemma. Suppose a kidnapper gets cold feet and wishes he had not taken his victim. He proposes to release her but only if she agrees not to give evidence against him. Yet he knows if he lets her go, she will be grateful, but she will have no reason by then not to break her promise and go straight to the police. She will be out of his power. So she assures him that she will do no such thing, but her assurances carry no conviction to the kidnapper, because he knows they are not worth the air they are spoken in; to go back on them will cost her nothing. The dilemma is really hers, not his. How can she commit herself to her side of the bargain? How can she make it costly for herself to break the deal?

She cannot. Schelling suggested that she should in some way compromise herself, by revealing a terrible crime she has committed in the past so that the kidnapper could be witness against her, and mutual deterrence would ensure that the deal sticks. But how many kidnappers' victims have something as awful as kidnapping to confess to? It is not a realistic solution to the dilemma, which remains insoluble for lack of enforceable commitment.

In real life, commitment problems are, however, more soluble, for an intriguing reason. We use our emotions to make credible commitments for us. Consider two of the examples Frank gives of such problems. First, two friends consider starting a restaurant, one cooking the food, the other keeping the books; each could easily cheat the other. The cook could exaggerate the cost of food; the accountant could cook the books. Second, a farmer must deter his neighbour from letting cattle stray into his wheat; yet the threat of a lawsuit is not credible because the costs would outweigh the value of the damage done.

These are not esoteric or trivial problems; they are the kind of thing that faces all of us repeatedly throughout life. Yet in each case, a rational person would come out badly. The rational entrepreneur would not start the restaurant for fear of being cheated – or would herself cheat for fear that her equally rational partner was cheating, and would thereby ruin the business. The rational farmer would not be able to deter his rational neighbour from letting cattle into his wheat, because he would not waste money going to court.

To bring reason to such problems, and to assume that others would, is to lose the opportunities they represent. Rational people would be unable to convince each other of their commitment and would never close the deals. But we don't bring reason to such problems; we bring irrational commitment driven by our emotions. The entrepreneur does not cheat for fear of shame or guilt, and she trusts her partner, knowing her to be a woman who does not like to face shame or guilt herself – a person of honour. The farmer fences in his cattle knowing that his neighbour's rage and obstinacy will cause him to sue even if it means ruining himself in the process.

In this way emotions alter the rewards of commitment problems, bringing forward to the present distant costs that would not have arisen in the rational calculation. Rage deters transgressors; guilt makes cheating painful for the cheat; envy represents self-interest; contempt earns respect; shame punishes; compassion elicits reciprocal compassion.

And love commits us to a relationship. Although love may not last, it is by definition a more durable thing than lust. Without love, there would be a permanent and shifting cast of sexual partners none of whom could ever elicit commitment to the bond. If you do not believe me, ask chimpanzees or their close relatives, bonobos, for this neatly describes their sex lives.

A few years ago, Dutch researchers discovered that if the male of a pair of small birds called blue tits is wounded by a sparrowhawk during egg laying, its mate will promptly seek another male to mate with. This is rational; the wounded male may die or pine away, and the female would be better off with another male. In order to interest another male in rearing her brood, she should give him a share of their paternity. But to human ears, the female's behaviour is almost unbelievably callous and heartless, however sensible it is. Likewise, I noticed when studying animals how lacking they usually are in a sense of grudge. They do not nurture thoughts of revenge on those that have harmed them; they simply get on with life. This is sensible, but it does mean that an animal can harm another without considering the consequences. Complicated emotions, so characteristic of human beings, prevent us deserting wounded mates or forgiving unfair slights. This, in the long run, is to our advantage, for it allows us to keep marriages together in bad times, or warn off potential opportunists. Our emotions are, as Frank has put it, guarantees of our commitment.[11]

Fairness matters

In his original paper on reciprocal altruism, Robert Trivers came up with much the same idea: emotions mediate between our inner calculator and our outer behaviour. Emotions elicit reciprocity in our species, and they direct us towards altruism when it might, in the long run, pay. We like people who are altruistic towards us and are altruistic towards people who like us. Trivers noticed that moralistic aggression serves to police fairness in reciprocal exchanges – people seem to be inordinately upset by 'unfair' behaviour. Likewise, the emotions of gratitude and sympathy are surprisingly calculating. Psychological experiments reveal – as experience confirms – that people are much more grateful for acts of kindness that cost the donor some large effort or inconvenience than for easy acts, even if the benefit received is the same. We all know the feeling of resentment at an unsolicited act of generosity whose intent is not to do a kindness but to make us feel the need to do a kindness in return. The emotion of guilt, Trivers argued, is used to repair relationships once the guilty person's cheating has been exposed. People are more likely to make altruistic reparative gestures out of guilt when their cheating has become known to others. All in all, the human emotions looked to Trivers like the highly polished toolkit of a reciprocating social creature.[12]

But whereas Trivers couched his version of the theory in terms of immediate reward through reciprocity, Frank reckons the commitment model rescues the altruism question from the clutches of such cynics. It does not try to take the altruism out of altruism. In contrast to explanations based on reciprocity and nepotism, the commitment model allows that genuine altruism can evolve.

> The honest individual in the commitment model is someone who values trustworthiness for its own sake. That he might receive a material payoff for such behaviour is beyond his concern. And it is precisely because he has this attitude that he can be trusted in situations where his behaviour cannot be monitored. Trustworthiness, provided it is recognizable, creates valuable opportunities that would not otherwise be available.[13]

To this a cynic might reasonably reply that the reputation for trustworthiness that honesty earns is itself just reward amply balancing the costs of occasional altruism. So, in a sense, the commitment model does take the altruism out of altruism by making altruism into an investment – an investment in a stock called trustworthiness that later pays handsome dividends in others' generosity. This is Trivers's point.

Therefore, far from being truly altruistic, the cooperative person is merely looking to his long-term self-interest, rather than the short term. Far from dethroning the rational man beloved of classical economists, Frank is merely redefining him in a more realistic way. Amartya Sen has called the caricature of the short-sighted self-interested person a 'rational fool'. If the rational fool turns out to be taking short-sighted decisions then he is not being rational, just short-sighted. He is indeed a fool who fails to consider the effect of his actions on others.[14]

However, such quibbling aside, Frank's insight is still remarkable. At its core lies the idea that acts of genuine goodness are the price we pay for having moral sentiments – those sentiments being valuable because of the opportunities they open in other circumstances. So when somebody votes (an irrational thing to do, given the chances of affecting the outcome), tips a waiter in a restaurant she will never revisit, gives an anonymous donation to charity or flies to Rwanda to bathe sick orphans in a refugee camp, she is not, even in the long run, being selfish or rational. She is simply prey to sentiments that are designed for another purpose: to elicit trust by demonstrating a capacity for altruism. This is not really an alternative interpretation from that proposed in the last chapter – that people do good deeds in order to win prestige that, through indirect reciprocity, they can later cash as a more practical good. Richard Alexander takes the philosopher Peter Singer to task for arguing that the existence of national blood banks that rely on generosity proves people are not motivated by reciprocity. It is true that people give blood in Britain in no expectation that they will be paid or will get preferential treatment if they need blood themselves. You get a cup of weak tea and a polite thank-you. But, says Alexander, 'Who among us is not a little humble in the presence of someone who has casually noted that he just came back from "giving blood"?'[5] People are not generally very secretive about blood donation. Giving blood and working in Rwanda both enhance your reputation for virtue and therefore make people more likely to trust you in prisoner's dilemmas. They scream out 'I am an altruist; trust me.'

The point, then, of moral sentiments in a situation resembling a prisoner's dilemma, is to enable us to pick the right partner to play the game with. The prisoner's dilemma is a dilemma only if you have no idea whether you can trust your accomplice. In most real situations, you have a very good idea how far you can trust somebody. Imagine, says Frank, that you have left £1,000 in an envelope with your name and address on it in a crowded theatre. Of all those people whom you know, are there some who you think would be more likely to return the envelope if they found it? Of course there are. So you distinguish among your acquaintances according to how much you can trust them to cooperate with you even in a situation in which they could get away undetected with not cooperating.

Indeed, as Frank has shown in his own experiments, if people are asked to play the prisoner's dilemma with each of a group of strangers in turn, but given just thirty minutes to meet the partners first, they prove remarkably good at predicting which of the strangers will defect and which will cooperate in the game (see Chapter Five). Consider, for example, how important a smile is from somebody you are meeting for the first time. It is a hint that this person desires to trust and be trusted; it could be a lie, of course, though plenty of people would bet that they could distinguish a fake smile from a 'real' one. Still harder is it to laugh convincingly if you are not amused, and in many people a blush is wholly involuntary. So our faces and our actions seem to advertise with disarming frankness just what is going on in our heads, which seems thoroughly disloyal of them. Dishonesty is so physiological it can be detected by a machine: a lie detector. Anger, fear, guilt, surprise, disgust, contempt, sadness, grief, happiness – all are universally recognizable, not just in one culture, but across the globe.

Such easily detected emotions plainly benefit the species in that they allow trust to go to work in society, but what possible use are they to the individual? Go back to the prisoner's dilemma tournaments of Chapter Three and recall how, in the world of defectors, a Tit-for-tat stratagem cannot take hold unless it finds other cooperators. Likewise, says Frank, in a world of people who find it easy to deceive themselves and their facial muscles – who are good at lying – a poor self-deceiver would suffer. But once he could find another poor self-deceiver, the two would hit it off. They would be able to trust each other and avoid playing the game with anybody else. To identify people who are not opportunists is an advantage; to be identified as a non-opportunist is equally an advantage for it attracts others of the same stamp. Honesty really is the best policy for the emotions.

One of Frank's strongest examples is the issue of fairness. Consider the game known as the 'ultimatum bargaining game'. Adam is given £100 in cash and told to share it with Bob. Adam must say how much he intends to give to Bob and, if Bob refuses the offer, neither will get anything at all. If Bob accepts, then he gets what Adam has offered. The logical thing for Adam to do, assuming he thinks Bob is also a rational fellow, is to offer Bob a derisory sum, say £1, and keep the remaining £99. Bob should rationally accept this, because then he is £1 better off. If he refuses, he will get nothing.

But not only do very few people offer such a small sum when asked to act Adam's part, even fewer accept such exiguous offers when playing Bob's part. By far the commonest offer made by real Adams is £50. Like so many games in psychology, the purpose of the ultimatum bargaining game is to reveal how irrational we are and wonder at the fact. But Frank's theory has little difficulty explaining this 'irrationality', even finding it to be sensible. People care about fairness as well as self-interest. They do not expect to be offered such a derisory sum by someone in Adam's position and they refuse it because irrational obstinacy is a good way of telling people so. Likewise, when playing Adam, they make a 'fair' offer of 50:50 to show how fair and trustworthy they are should future opportunities arise that depend on trust. Would you risk your good reputation with your friends for a lousy £50?

But this is the reasoning of reciprocity, not fairness. The economist Vernon Smith has subtly varied the ultimatum bargaining game to reveal that it does not say much about an innate sense of fairness, but instead supports the argument that reciprocity motivates people. If, among a group of students, the right to play Adam is 'earned' by scoring in the top half of the class on a general knowledge test, then Adams tend to be less generous. If the rules are changed so that Bob must accept the offer – which Smith calls the 'dictator game' – then once more the offers are less generous. If the experiment is presented not as an ultimatum given by Adam, but as a transaction between a buyer and a seller in which Bob must quote a price, then again Adams are less generous. And if the experiment is conducted in such a way as to protect the anonymity of Adam, then again Adams are less generous. Now, with their identity protected from the experimenter, seventy per cent of Adams offer nothing in the dictator game. It is as if the subjects think the experimenter will not ask them back (the experimental sessions are profitable) unless they display pro-social behaviour.

In all these new circumstances, people should be just as generous if it is an innate sense of fairness that motivates them. Yet they are not. They reveal a strict sense of opportunism instead. So why are they generous in the original game? Because, argues Smith, they are obsessed with reciprocity. Even when the game is only played once, they are concerned to protect their personal reputation for being somebody who can be trusted not to be too nakedly opportunistic at others' expense.[16]

Smith uses a game called the 'centipede game' to ram home the message. In this game Adam and Bob have the chance to pass or take the money on each turn. The longer they pass, the more money there is, but eventually the game hits the end and Adam gets the money. So Bob should reason with himself that he should not pass on his last go; then Adam should reason that Bob will do this so he should not pass on his penultimate go; and so on until each is led to the conclusion that he should stop the game at the first chance.

Yet people do not. They routinely allow others to win lots of money by passing. The reason, clearly, is that they are trading – rewarding the other person for not being selfish, hoping for reciprocal generosity when their turn comes. Yet there is no systematic switching of roles.

Robert Frank's commitment model is in some ways a rather old-fashioned idea. What he is saying is that morality and other emotional habits pay. The more you behave in selfless and generous ways the more you can reap the benefits of cooperative endeavour from society. You get more from life if you irrationally forgo opportunism. The subtle message of both neo-classical economics and neo-Darwinian natural selection – that rational self-interest rules the world and explains people's behaviour – is inadequate and normatively dangerous. Says Frank:

> [Adam] Smith's carrot and Darwin's stick have by now rendered character development an all but completely forgotten theme in many industrialized countries.[17]

Tell your children to be good, not because it is costly and superior, but because in the long run it pays.

The moral sense

Robert Frank is an economist, but his ideas echo and are echoed by the writings of two psychologists. Jerome Kagan is a child psychologist whose studies of the inheritance, development and causes of personality lead him inexorably to emphasize emotion rather than reason as the wellspring of human motivation. The desire to escape or avoid guilt, says Kagan, is a human universal, common to all people in all cultures. The kinds of events that cause guilt may vary from culture to culture – being unpunctual, for instance, is a very Western thing to feel guilty for – but the reaction to guilt is the same the world over. Morality requires an innate capacity for guilt and empathy, something children of two years old clearly lack. Like most innate capacities (language, say, or good humour), though, the moral one can be nurtured or suppressed by different kinds of upbringing; so to say that the emotions that fuel morality are innate is not to say they are immutable.

Kagan's theory of childhood morality is therefore like Frank's commitment model in its emphasis on irrational emotions.

> Construction of a persuasive basis for behaving morally has been the problem on which most moral philosophers have stubbed their toes. I believe they will continue to do so until they recognize what Chinese philosophers have known for a long time: namely, that feeling, not logic, sustains the superego.[18]

Incidentally, it seems as if vervet monkeys, like two-year-olds, completely lack the capacity for empathy. If one vervet monkey makes an alarm call, it does not cease merely because another is already calling so must already be aware of the danger. Vervet monkeys never correct their babies' mistakes in making alarm calls. And vervets do not make alarm calls when a baboon approaches. Baboons eat baby vervets but not adults. Thus the monkey's alarm is sublimely self-centred. As Dorothy Cheney put it after studying both vervets and baboons, 'Signalers do not recognize the mental state of listeners, so they cannot communicate with the intent of appeasing those who are anxious or informing those who are ignorant.'[19] They cannot empathize. This is such an obvious difference between human beings and other animals that it is hard to step back and see it for the idiosyncrasy that it is. We do not cut into queues, because we care what other people – even strangers – think of us. Other animals do not.

A decade after Kagan's book was published and six years after Frank's, James Q. Wilson published *The Moral Sense*, which makes many of the same arguments from a criminologist's perspective. 'What most needed explanation, it seemed to me, was not why some people are criminals but why most people are not.' Wilson chides philosophers for not taking seriously the notion that morality resides in the senses as a purposive set of instincts. They mostly view morality as merely a set of utilitarian or arbitrary preferences and conventions laid upon people by society. Wilson argues that morality is no more a convention than other sentiments such as lust or greed. When a person is disgusted by injustice or cruelty he is drawing upon an instinct, not rationally considering the utility of the sentiment, let alone simply regurgitating a fashionable convention.

For example, even if you dismiss charitable giving as ultimately selfish – saying that people only give to charity in order to enhance their reputations – you still do not solve the problem because you then have to explain why it does enhance their reputations. Why do other people applaud charitable activity? We are immersed so deeply in a sea of moral assumptions that it takes an effort to imagine a world without them. A world without obligations to reciprocate, deal fairly, and trust other people would be simply inconceivable.[20]

Psychologists, therefore, are converging with Robert Frank's economic argument that emotions are mental devices for guaranteeing commitment. But perhaps the most remarkable convergence comes from the study of broken brains. There is a small part of the prefrontal lobe of the human brain, which, when damaged, turns you into a rational fool. People who have lost that part of their brain are superficially normal. They suffer no paralysis, no speech defect, no loss in their senses, no diminution in their memory or general intelligence. They do just as well in psychological tests as they did before their accidents. Yet their lives fall apart for reasons that seem more psychiatric than neurological (oh false dichotomy!). They fail to hold down jobs, lose their inhibitions, and become paralytically indecisive.

But this is not all that happens to them. They also literally lose their emotions. They greet misfortune, joyful news and infuriating checks with equanimity and reason. They are simply flat, emotionally.

Antonio Damasio, who described these symptoms from twelve patients in his book *Descartes's Error*, thinks it is no accident that decision making and emotion go together. His patients become so cold-blooded about rationally weighting all the facts before them that they cannot make up their minds. 'Reduction in emotion may constitute an equally important source of irrational behavior,' he speculates.[21]

In short, if you lack all emotions, you are a rational fool. Damasio makes this case without apparently knowing that economists like Robert Frank, biologists like Robert Trivers and psychologists like Jerome Kagan have come to similar conclusions from different evidence. It is a remarkable coincidence.

Patience is a virtue, a virtue is a grace, and Grace is a little girl who wouldn't wash her face. This meaningless little ditty now seems to contain a gem of an insight that summarizes the commitment model's main discovery. Virtue is indeed a grace – or an instinct as we might put it in these less Augustinian days. It is something to be taken for granted, drawn on and cherished. It is not something we must struggle to create against the grain of human nature – as it would be if we were pigeons, say, or rats with no social machine to oil. It is the instinctive and useful lubricant that is part of our natures. So instead of trying to arrange human institutions in such a way as to reduce human selfishness, perhaps we should be arranging them in such a way as to bring out human virtue.

Let others be altruists

There is a paradox in the common view of self-interest. People are generally against it; they despise greed and warn each other against people who have a reputation for too closely pursuing their own ambitions. Similarly, they admire the disinterested altruist; tales of such people's selflessness become legend. So it is pretty clear that on a moral level, everybody agrees that altruism is good and selfishness bad.

So why are more people not altruists? The exceptions – the Mother

Theresas and saints – are almost by definition remarkable and rare. How many people do you know who are true altruists, always thinking about others and never themselves? Very, very few. Indeed, what would you say to somebody close to you who was being truly selfless – a child, say, or a close friend, who was continually turning the other cheek, doing little tasks at work that others should have done, working for no reward in a hospital emergency room, or giving his weekly pay to charity? If he did it occasionally, you would praise him. But if he did it every week, year after year, you would start to question it. In the nicest way you might hint that he should look out for himself a little more, be just a touch more selfish.

My point is that while we universally admire and praise selflessness, we do not expect it to rule our lives or those of our close friends. We simply do not practise what we preach. This is perfectly rational, of course. The more other people practise altruism, the better for us, but the more we and our kin pursue self-interest, the better for us. That is the prisoner's dilemma. Also, the more we posture in favour of altruism, the better for us.

I believe this explains the general mistrust in which both economics and selfish-gene biology is held. Both disciplines claim repeatedly and with little effect that they are being misunderstood; they are not recommending selfishness, they are recognizing it. It is only realistic, economists say, to expect human beings to react to incentives with a view of their self-interest – not just or good, but realistic. Likewise, say biologists, it is plausible to expect genes to show an evolved ability to do things that enhance the chances of their own replication. But we tend to see it as a bit naughty to take this view; somehow not politically correct. Richard Dawkins, who coined the phrase 'selfish gene', says that he drew attention to the inherent selfishness of genes not to justify it, but the reverse: to alert us to it so we can be aware of the need to overcome it. He urged us to 'rebel against the tyranny of the selfish replicators'.[22]

If the commitment model is right, though, the critics of selfish schools have a point, for everything becomes normative. If people are not rational maximizers of self-interest, then to teach them that such behaviour would be logical is to corrupt them. Indeed, this is just what Robert Frank and many others have found: those students who have been taught the nostrums of neo-classical economics are much more likely to defect in prisoner's dilemma games than, for instance, astronomy students.

The virtues of tolerance, compassion and justice are not policies towards which we strive, knowing the difficulties upon the way, but commitments we make and expect others to make – gods we pursue. Those who raise difficulties, such as economists saying that self-interest is our principal motivation, are to be distrusted for their motives in not worshipping the gods of virtue. That they do so suggests that they may not themselves be believers. They show, as it were, an unhealthy interest in the subject of self-interest.

Theories of moral sentiments

Frank's 'theory of moral sentiments' fleshes out Adam Smith's, first advanced in his book of that name published in 1759. It also begins to build a bridge between Smith's apparently irrational assumption that people are driven by moral sentiments and his devotion to rational self-interest as the wellspring of a successful economy: a bridge between his first and second books.

In his first book Adam Smith argued that if individuals had sufficient common interest in the good of their group, they would combine to suppress the activities of members acting contrary to the group's welfare. Bystanders would interfere to punish anti-social actions. But in his second book, Smith seemed to undermine this argument by suggesting that societies are not public goods carefully protected by individuals but are the almost inevitable side-effect of individuals striving in their own, individual interests.

The Germans, who, it seems, in their methodical manner commonly read both *Theory of Moral Sentiments* and the *Wealth of Nations*, have coined a pretty term, *Das Adam Smith Problem*, to denote the failure to understand either which results from the attempt to use the one in the interpretation of the other.[23]

Frank's theory of moral sentiments resolves this paradox and builds another, more modern bridge – between reciprocity and groupishness. By emphasizing that the challenge in the prisoner's dilemma game is to attract the right partner, he shows how reciprocators precipitate out of society, leaving the selfish rationalists to their fate. The virtuous are virtuous for no other reason than that it enables them to join forces with others who are virtuous, to mutual benefit. And once cooperators segregate themselves off from the rest of society a wholly new force of evolution can come into play: one that pits groups against each other, rather than individuals.

CHAPTER EIGHT

The Tribal Primates

In which animals cooperate in order to compete

Brute creatures are impressed and actuated by various instincts and propensions; so also are we ... The fact then appears to be, that we are constituted so as to condemn falsehood, unprovoked violence, injustice, and to approve of benevolence to some preferably to others.

Bishop Joseph Butler, *Of the Nature of Virtue*, 1737

Imagine, if you can, that you are a male baboon on a plain in East Africa. Now this will take some doing because baboon society is rather strange in many ways. But to help you learn the ropes, I will give you an important hint about what your fellow baboons frequently get up to. They combine into coalitions with the intention of stealing other males' mates. So if you are in the happy situation of sitting alongside a female baboon who is in season, enjoying a quiet honeymoon with her, and you see another male baboon walking up to his friend with a particular movement known as head-flagging, watch out. The head-flagging baboon is saying to his friend, 'How about joining me to attack that guy over there and steal his girl?' Two against one is a foregone conclusion and you are soon running for your life across the savanna with a sore behind.

In baboon society, this is the way junior males get to have sex: they gang up on seniors and drive them away from their monopolized females. But only one of the two coalition partners actually has sex; the other merely takes part in the fight for nothing. So why does he do it? Is he an altruist? The first answer, supplied by the zoologist Craig Packer in 1977, was that he does it because he expects to be rewarded by a similar favour from the animal he helped on this occasion. Thus, it is the baboon who asks for the favour – the first head-flagger – who gets to have sex, but he commits himself to return the favour if asked in the future, just like Wilkinson's vampire bats.[1]

Indeed, it was after a visit to Africa to watch baboons that Robert Trivers wrote his seminal theory of reciprocal altruism, and it was to test Trivers's theory that Packer did his study a few years later. Baboons, it seems, are the original archetypes of reciprocating altruists: the typical Tit-for-tat players.

The only trouble is, Packer was wrong. When other scientists looked at baboons for longer they found that it is very much not a foregone conclusion who gets the girl. Indeed, there is an unseemly chase between the coalition partners to catch her once the previous consort is beaten off. So there is nothing altruistic here at all; just self-interest. Baboon A's only hope of having sex is to join forces with B and attack C to steal his female, then hope he gets to her before B does. Both A and B get an immediate benefit from cooperating: a fifty per cent chance of having sex.

In any case, the baboon's situation is not a prisoner's dilemma, because there is no temptation to defect. If A and B refuse to form a coalition, far from one benefiting greatly, they both suffer: neither stands a chance of winning a female.[2]

None the less, whether the baboons are playing Tit-for-tat or not, they are still cooperating and thus discovering the virtues of cooperation. They are joining forces to achieve an end. Two weak individuals, by cooperating, can beat a stronger one. What counts is not strength but social skills. Brute force is tamed by virtue. The well-connected will inherit the earth. Is this a first, primitive step on the ladder of primate cooperation that led to human society? If so, it would hardly have pleased Prince Kropotkin, for the purpose of cooperation is not a noble, communal goal – the good of baboon society – but a narrowly selfish end: sexual monopoly irrespective of the wishes of the female concerned, let alone those of her previous consort. Cooperation was first used, not for virtuous reasons, but as a tool to achieve selfish results. And if we are to celebrate the unusually cooperative nature of our societies, we must first recognize the base metal from which it was forged.

Baboons are not alone in this. Throughout the societies of monkeys, cooperation is encountered almost exclusively in the context of competition and aggression. It is, in male monkeys, a way of winning fights. If you wish to see monkeys cooperating in coalitions and alliances, your best bet is to catch them fighting each other. Colobus monkeys steal harems of females from each other by attacking harem owners with the help of male friends.[3]

The baboon story does at least have a simple lesson which might serve you in good stead if you are ever reincarnated as a baboon. You will know that coalitions serve the purpose of sexual kidnap. But suppose you do not come back as a baboon but as a bonnet macaque, in many ways a rather similar animal: a ground-dwelling monkey, fairly strong and fierce as monkeys go, and living in large, hierarchical societies, just like baboons.

In one respect, however, life among the bonnet macaques is most unlike that among the baboons. In baboons, coalitions are few, occasional and stable. A and B are best friends and very occasionally they gang up to grab females belonging to some other baboon. Baboon fights are mostly between one animal and another. Male bonnet macaques, on the other hand, fight frequently among themselves and most fights are between 'teams' of two animals, rather than one-against-one. In bonnet macaques, coalitions are everywhere. On average one is formed every thirty-nine minutes. Every male in the troop will at some time form a coalition with every other male. Male bonding is not confined to the odd head-flagging precursor to a battle; it is the stuff of life. Males groom each other, play with each other, huddle together, snooze in each other's arms, wander about in pairs and generally spend vast amounts of effort creating and maintaining temporary friendships with each other. These coalitions are usually instigated by a fight, and usually consist of some monkey coming to the aid of the animal that started the fight. But a few hours later the instigator could find himself facing his erstwhile ally now in a coalition with some other male. It is all very bewildering.

But it is not random. On average, males support males that have supported them or groomed them in the past and, on average, rank plays a big role: supporters in fights are usually senior males coming to the aid of junior ones. Junior ones return the compliment by grooming their senior allies. Unlike baboons, the coalitions are negative as well as positive: male bonnet macaques take revenge on those that have helped their enemies as well as coming to the aid of those they have received help from.

The world of the male bonnet macaque, in other words, is one of shifting, continuous and frequent friendships, repaid favours, alliances and loyalties. It takes up a lot of their time. So what is it all about?

Joan Silk, who has been studying the species for many years in a captive troop maintained in California, has not got the foggiest idea. Coalitions do not help males win females as they do for baboons; they do not alter the pecking order as they can for chimpanzees; they do seem to help males win fights against each other but, since erstwhile friends can become enemies, any advantage is temporary. Silk remains genuinely baffled. If any reader does get reincarnated as a bonnet macaque, perhaps he can send Joan Silk a postcard telling her what it is all about.[4]

Monkeys with attitude

Silk and her ilk study monkeys not just because monkeys are interesting in themselves, but because they are related to us, albeit more distantly than apes. The burgeoning of primatology in the 1970s and 1980s laid bare a plethora of sophisticated social set-ups throughout the family to which humankind belongs. Anybody who thinks this is irrelevant to the study of human beings must be a Martian. We are primates, and we can learn about our roots by studying our relatives.

This premise can lead quickly to two fallacies. The first is that primatologists are somehow claiming that human beings are the same as monkeys in every respect and detail, which is clearly nonsense. Each monkey and each ape has its own social system, unique to that species; but there are still common threads. Each species of monkey looks different from each other species yet it still makes sense to say that all species of monkey look rather like each other, in comparison, say with all deer. Likewise, all primate species behave in different ways, but in ways that are recognizably primate-like.

The second fallacy is to suppose that monkeys are somehow more primitive than people socially. Monkeys are not our ancestors any more than we are theirs. We share a common ancestor with all monkeys, but we have altered the body plan and the social habits of that ancestor in idiosyncratic ways. So has each species of monkey.

Deriving lessons from nature is a tricky feat. You must steer your craft carefully between two terrifying temptations. On one side, Scylla cries out to you to look for direct animal parallels, ways in which we are just like our cousins. Thus Kropotkin argued that because ants were nice to each other, so must we be instinctively virtuous. Thus Spencer argued that because nature is a pitiless struggle, pitiless struggles must be virtuous. But we are not like animals in every respect. We are unique, we are different, just as every species is unique and different from every other; biology is a science of exceptions, not rules; of diversity, not grand unified theories. That ants are communitarian says nothing about whether man is virtuous. That natural selection is cruel says nothing about whether cruelty is moral.

Yet beware of steering your craft too far the other way. Charybdis cries seductively from that side to emphasize human uniqueness. Nothing, she says, can be learnt from nature. We are ourselves, in the image of god or of culture (depending on taste). We have sex drives because we are taught to have them, not because of instinct. We speak languages because we teach each other to speak. We are conscious, rational and free-willed, not like those inferior things called animals. Virtually every high priest of the humanities, of anthropology and of psychology preaches the same old, defensive sermon of human uniqueness that theologians clung to when Darwin first shook their tree. Where Richard Owen sought then desperately for proof in the hardware of the human brain for an object that was unique to humankind – and believed he had found it in the hippocampus minor, an odd little bump on the brain – so today anthropologists demand that the existence of culture, reason or language exempts us from biology.

The last bastion of this argument is that even if human beings have evolved natures, one can never be sure that one is seeing their instincts in action, rather than their conscious or cultural decisions. Wealthy people favour their sons rather than their daughters, as do many primates that find themselves high in social rank. But this need not be a shared instinct between human beings and monkeys. It might be that people have rediscovered the same logic by consciously deducing that sons can use wealth as a passport to greater reproductive success than daughters. For human beings, you can never entirely reject the culture hypothesis. As Dan Dennett put it, in *Darwin's Dangerous Idea*, 'If a trick is that good, then it will be rationally rediscovered by every culture without need of genetic descent.'[5]

But this argument cuts both ways and inflicts a sharper wound on the orthodoxy of the environmental determinists than they realize. For every time you see human beings behaving adaptively, you may think you are seeing conscious or cultural decisions, but you might just be seeing evolved instincts. Language, for instance, looks like a cultural artefact – after all it varies between cultures. But to speak enthusiastically, grammatically and with a large vocabulary is pre-eminently an instinct of our species that cannot be taught, only learnt.[6]

The study of animals has profound implications for our understanding of the human mind – and vice versa. As Helena Cronin has argued, 'to erect a biological apartheid of "us" and "them" is to cut ourselves off from a potentially useful source of explanatory principles . . . Admittedly we're unique. But there's nothing unique about being unique. Every species is in its own way.'[7] That we now know how the complex societies of monkeys and apes work is highly relevant to understanding our own society. An evolutionary perspective inevitably eluded Hobbes and Rousseau; less forgivably it still eludes some of their intellectual descendants. The philosopher John Rawls asks us to imagine how rational beings would come together and create a society from nothing, just as Rousseau imagined a solitary and self-sufficient proto-human. These were only thought-experiments, but they serve to remind us that there never was a 'before' society. Human society is derived from the society of *Homo erectus*, which is derived from the society of *Australopithecus*, which is derived from the society of a long-extinct missing link between humans and chimps, which in turn was derived from the society of the missing link between apes and monkeys, and so on, back to an eventual beginning as some sort of shrew-like animals that perhaps genuinely lived in Rousseauian solitude. Of course, we cannot go back and examine the societies of *Australopithecus*, but we can make some informed guesses based on anatomy and on modern parallels.

First, we can say that our ancestors were social. All primates are, even the semi-solitary orang-utans. Second, we can say that there was a hierarchy within each group, a pecking order; that this hierarchy was more marked among males than females – these facts are true of all primates. But we can then say something rather interesting, albeit with less confidence: our ancestors' hierarchies were less rigid and more egalitarian than those of monkeys. This is because we are apes, and cousins of chimpanzees in particular.

In monkeys, despite the invention of cooperation, weak and junior male monkeys still occupy lower rank and mate with fewer females than strong and senior ones. Brute force may not be as reliable as it is among sheep and elephant seals, but it is still highly influential. In the societies of chimpanzees, however, the importance of physical prowess is markedly less. The top male chimpanzee in a troop is not necessarily the strongest; instead, it is usually the one best at manipulating social coalitions to his advantage.

In the Mahale mountains in Tanzania lives a powerful alpha male chimpanzee called Ntogi, who frequently catches monkeys or antelope for food. He shares the meat with his mother and his current girlfriends, as is normal (Chapter Five), but he also carefully supplies some males with meat as well. He gives it to middle-ranking males and older males. He never gives meat to young males or to senior males. In other words, like a good client of Machiavelli, he cultivates his best constituency: the middle-management males on whom he relies to form coalitions against the ambitious young and his immediate rivals. The meat is the currency in which he pays his allies to keep him in power.[8]

Unlike baboons, which form coalitions specially to steal females from other males higher in the hierarchy than themselves, chimps use coalitions to modify the social hierarchy itself. This has been observed in wild chimps in Tanzania, but the best-documented case comes from a group of chimpanzees living on a small island in a lake at Arnhem zoo, which were closely studied by Frans de Waal in the 1970s and 1980s.

In 1976, a powerful chimp called Luit became the dominant alpha male in the group by dominating the previous alpha, Yeroen. Before this Luit had tended to cultivate other males who had just won fights, joining in the attacks on the loser. But once he became alpha, he switched to supporting losers, taking the side of the underdog and so stopping a fight. There was nothing unselfish in this, de Waal reckons, just a careful expression of self-interest. Luit was cultivating his grass-roots support and keeping on top of any potential rivals in just the same way as any medieval king or Roman emperor. Luit was especially popular with the females, on whom he could rely for support in a tight spot.

However, Luit was soon toppled from the alpha position by a conspiracy between his predecessor and his successor. Yeroen, the older male Luit had toppled, formed a coalition with Nikkie, an ambitious young chimp who was not as strong as Luit on his own. The two attacked Luit and after a savage fight, deposed him. Nikkie became the alpha male, though he had to rely on Yeroen's support in any fight, particularly one involving Luit. It looked as if Nikkie had manipulated a cooperative relationship to his benefit.

But Yeroen was the most cunning of the three. He set about translating his new position as power behind the throne into sexual success, and was soon the most sexually active male in the group, performing nearly forty per cent of all matings. He did this by playing on Nikkie's need for his support. In return for his help when Nikkie asked for it, he would demand Nikkie's support in removing Luit if Luit paid too much attention to a fertile female, and would then mate with the female himself. In de Waal's interpretation, Nikkie and Yeroen had a deal: Nikkie could have the power if Yeroen had a large share of the sex.

So, when Nikkie started to renege on the deal, he got into trouble. Nikkie began to do more of the mating himself, and Yeroen was soon spending half as much time copulating as before. This Nikkie achieved by ceasing to intervene against Luit, leaving Yeroen to fight his own battles against him. Nikkie now used either Yeroen or Luit to help him achieve his aims in contests with other apes. He was dividing and ruling with increasing confidence. One day in 1980, however, he went too far. Nikkie and Luit together displaced Yeroen from a female several times, and Nikkie then failed to respond to Yeroen's screamed request that he stop Luit climbing a tree after a fertile female – at which an enraged Yeroen attacked Nikkie. Yeroen, it seemed, had had enough of Nikkie's rule. A few days later, after a fierce fight during the night that left both Yeroen and Nikkie injured, Nikkie was no longer alpha male. Luit was back in power.[9]

Shortly after I first read the story of the Arnhem chimpanzees, I happened to be reading an account of the Wars of the Roses. Something nagged at the back of my mind. The tale was uncannily familiar, as if I had just read it in another form. Then it dawned on me. Margaret of Anjou, the queen of England, was Luit. Edward IV, the usurper son of the Duke of York, was Nikkie, and the wealthy earl known as Warwick the Kingmaker was Yeroen. Consider: with Warwick's help, the Duke of York toppled the incompetent and hen-pecked Henry VI. After York was killed, his son Edward IV became king, but nervous of Warwick's power allowed his wife's family to build up a rival faction at court to undermine Warwick. An increasingly disenchanted Warwick formed an alliance with Henry VI's wife, Margaret of Anjou, drove Edward into exile and seized back the throne for his new puppet, the bewildered Henry VI. But Edward successfully fomented rebellion against Warwick, killed him in battle, captured London and had Henry VI murdered. It is almost exactly the same story as Luit, Nikkie and Yeroen. At Arnhem, Luit, too, was eventually killed, by Yeroen.

The story of the Arnhem chimpanzee politics illustrates two central themes of chimpanzee life. The first is that the relations within these coalitions seem to be reciprocal. Unlike in monkeys, a coalition is a strictly symmetrical relationship. If A intervenes on behalf of B, either to defend him when attacked or to support him when he starts a fight, B must later do so on behalf of A or the coalition will fall. The Arnhem chimps clearly play Tit-for-tat.

The second theme is that power and sexual success can be achieved by coalitions of weaker individuals over stronger ones, a process taken to even greater extremes in human beings, where politics in hunter-gatherer bands seem sometimes to consist of little else but the formation of subordinate coalitions to pre-empt dominant individuals wielding power. The theme of kings and chiefs reined in and dominated by coalitions of their individually weaker inferiors is a common one from all of history – from Frazer's *The Golden Bough*, through the consulship of the Roman republic all the way up to the American constitution. To neutralize the power of an alpha male requires a large coalition, larger than chimpanzees usually achieve.[10]

The dark side of the dolphin

It is no coincidence that baboons form coalitions and have relatively big brains; or that chimpanzees rely still more heavily on coalitions and have even bigger brains for their body size. To use cooperation as a weapon in social relations requires individuals to keep a record of who is an ally and who a foe, who owes a favour and who bears a grudge – and the more memory and brainpower available, the better the calculation can be done. It will not have escaped the reader's notice that there is another ape with an even bigger relative brain size. But the human being is not the only species on earth with a bigger brain relative to its body size than a chimpanzee. There is one other: the bottlenose dolphin.

Bottlenose dolphins are far brainier than other dolphins and whales, to about the same degree as human beings are brainier than other apes. If brainpower limits or evolves from cooperative skill, then it is among bottlenoses that we might expect to find even more cooperation. Dolphin sociology is in its infancy, but the early results are exciting because they reveal some seminal similarities with apes, but also some salutary differences.

The most studied group of bottlenose dolphins is a collection of several hundred that live in a shallow, clear-water bay called Shark Bay on the coast of Western Australia. Some of the dolphins have been coming to the beach to be fed fish by local tourists since the

1960s, which makes it easy to find and observe them. Richard Connor and his colleagues have now been studying them for ten years with astonishing results. Those who prefer to believe that dolphins are somehow mystically perfect, peaceable and generally holistic had better stop reading here or risk losing their treasured preconceptions!

The Shark Bay dolphins live in a 'fission-fusion society', superficially not unlike that of spider monkeys or chimpanzees. This simply means that all members of the social group are rarely or never together at one time: acquaintanceships overlap and friendships are fluid. But there is one exception to this flexible rule. Adult male dolphins travel in twos or threes, and each pair or triplet is a close alliance of two or three firm friends. By following three pairs and five triplets, Connor and his colleagues pieced together the purpose of these alliances.

When a female dolphin comes into season, a male alliance often 'kidnaps' her for some days from the group in which she lived. The males then swim with her, one on either side and one – 'the odd one out' – nearby. She sometimes tries to escape and occasionally succeeds by dashing off through the water. Nor are her suitors especially gentle with her. They chase her when she tries to escape, hit her with their tails, charge, bite and slam their bodies into her to keep her going where they want. They also indulge in spectacular displays of synchronized jumping, diving and swimming – just as trained dolphins will do in captivity. And they mate with her, apparently taking turns or even attempting to do it simultaneously.

There seems to be little doubt that the males are trying to monopolize the fertile female in order to father her next offspring, and that they do so in pairs or triplets for the obvious reason that a single individual could never control the movements of a female or keep her from being stolen by another male or pair of males. Likewise, since fatherhood is an indivisible resource, it makes sense that three is the upper limit of alliance size. Larger alliances would offer diminishing parental returns to the males even if they were more successful in herding females.

However, Connor's team then discovered that male alliances steal females from each other and they do so by forming 'second-order' alliances with other male coalitions. These allies are recruited especially for the occasion of the theft. For instance, Connor's team once saw a triplet called B come to the feeding beach, where they watched another triplet, H, which had a female in tow. B then left, swam a mile to the north and returned with a pair, A. The five dolphins proceeded to attack H and stole the female from it, at which A departed, leaving B in control of the female. A week later, B returned the favour by helping A to steal a female from H. A and B often help each other this way, as do H, G and D: the coalitions are affiliated with others into super-coalitions.[11]

This is exactly how baboons use allies – X recruits Y to steal a female from Z – except for two features. In dolphins, X, Y and Z are not individuals but teams of friends; and in dolphins there is no question of who is going to benefit from the theft of the female: one alliance is merely performing a selfless act of assistance. Indeed, the assisting alliance sometimes already has a female in tow (and may lose her in the commotion) when they help steal another for their allies – yet they can never control more than one female at a time. Far from helping the thieves out of self-interest, they are being immediately generous. Connor and his colleagues believe, but have not yet proven, that the relationship between a friendly pair of alliances is reciprocal. The dolphins therefore do something no primates except humans do: they form second-order alliances – coalitions of coalitions. In baboon and chimpanzee society all relationships between coalitions are competitive, not cooperative.

This leads to one of the most intriguing implications of the dolphin work. There is as yet no good evidence that dolphin societies are closed societies; that is, dolphins do not appear to divide up territorially into troops, or tribes or bands. Most primates do. A chimpanzee may live in a loose and fluid group and only occasionally see some of its compatriots, but it stays within that group's territory and it treats outsiders of the group as enemies. If it is a male, it will probably never leave the troop within which it was born, whereas females quite often leave their natal troops and join a different one. Baboons are the opposite. Males, once mature, leave the troop of their birth and force their way into another troop, usually at the top of the pecking order. This migration between groups prevents inbreeding.

Why is it males that leave in baboons and females in chimps? The reason may be the aggressive xenophobia displayed by male chimpanzees, which itself may be a consequence of the tendency of male chimpanzees to form coalitions. A lone male chimp, wandering into the territory of a neighbouring troop, faces almost certain death. Wherever they have been studied in East Africa, chimps have been found to practise something akin, if not to human warfare, then to raiding. A group of male chimpanzees sets out silently and purposively towards the territory of their neighbours. If they encounter a strong contingent of rival males, they retreat. If they encounter a female, they may try to bring her back to their territory. If they encounter a single male, they may attack it and kill it. One troop at Gombe studied by Jane Goodall in this manner exterminated the males in a small neighbouring troop and claimed all the females. Another troop in the Mahale mountains achieved the same result.

There is nothing strange in the animal kingdom about territoriality or even savage aggression between rival males. What is unusual (though not unique – wolves are another example) about the chimpanzees is the fact that the territory is defended by a group rather than an individual. Indeed, group territorial defence is nothing more than an extension of the coalition building that we witnessed between individuals such as Nikkie and Yeroen. Recall that Luit, when he became alpha male, supported losers against their persecutors. Alpha males also intervene to prevent fights happening at all. They have an important pacifying role. The reason, possibly, is to prevent the group breaking up, which in turn is important because larger groups are better able to resist the raids of their neighbours. When a group of males goes on a raid, the alpha behaves as if he must get the backing of his coalition partners before launching an attack. There was an occasion filmed at Gombe where the alpha Goblin apparently could not get the assent of some senior colleagues to pursue an action against some enemies, and the troop disengaged.

In chimps, therefore, the most important coalition of all is the one between all adult males of the same troop against all adult males of the enemy troop. This 'macro-coalition' only comes into play when danger threatens from 'abroad' or when it intends to threaten danger abroad itself. Male chimpanzees avoid the boundaries of their territories except when in fairly large groups; female chimpanzees stay away from such zones of danger altogether.

If it is true that bottlenose dolphins do not live in closed, territorial societies, then their coalitions of coalitions make perfect sense. A group of males cannot plausibly defend an area of sea against another group, or for that matter a group of females, so xenophobic hostility makes little sense. Even in clear water, a dolphin can escape detection from another just a mile or so away, especially if it remains silent – visibility is usually far better on land. So the purpose of dolphin coalitions is not to defend a group of females and a territory but to achieve occasional, temporary successes in herding individual females and to steal such females from other coalitions.[12]

The tribal age

Lethal inter-group violence is probably a characteristic we share with chimpanzees, as Richard Wrangham has argued. But we have brought something special to it: weapons. Once armed with a projectile weapon, such as a spear or even an accurately thrown rock, a man can attack other men with greater impunity. He need not risk injury himself if he has surprise on his side and his enemy is unarmed. This is very different from the danger even a group of chimpanzees face when they attack an enemy. The attackers could easily end up with broken bones, gashed skin or a missing eye. It takes on average twenty minutes for three or four chimpanzees to kill one other chimpanzee. Because of weapons, one human being can kill another with a single blow – and from a safe distance, too.

Projectile weapons were probably invented for hunting originally but, if so, there is something strange about them. Because they gradually increased the range at which a man could bring down an animal, they would in theory have made it less necessary for people to hunt in large groups, not more so. Armed with a bow and arrow a man can stalk his prey alone, whereas armed with rocks and clubs his best hope is that his allies will drive the game into an ambush.

The real significance of the invention of throwing weapons was that they made warfare more profitable and less risky. This would have increased the reward of joining a large coalition, for better defence and attack. It is perhaps no accident that *Homo erectus*, the first of our ancestral species to make sophisticated stone tools in large quantities, rapidly acquired a much larger stature and a thicker cranium. He was being regularly struck on the head. The relationship between weapons and coalitions was symbiotic. It has been obvious for years to anthropologists that weapons make dominance a chancy business, and thus require a leader to lead more by persuasion than by coercion. The !Kung people of South Africa have a habit of saying, during an argument: 'We are none of us big and others small; we are all men and can fight. I am going to get my arrows.' In his stories of Prohibition-era New York, Damon Runyon's slang for guns was 'equalizers'.[13]

Weapons are what make us different from chimpanzees and bottlenose dolphins. The shape of human society combines features from both the chimpanzee and the dolphin. Like chimps, we are xenophobic. All human preliterate societies, and all modern ones as well, tend to have an 'enemy', a concept of them and us. This effect is especially strong where human tribal societies consist of bands of related men and their wives and dependants – a common form of tribalism known as the fraternal-interest group. In other words, the more men stay in their native bands while women migrate, the more antagonism there is between groups. Matrilineal and matrilocal societies are a little less prone to feuding and warfare, in just the same way that matrilineal, matrilocal baboon societies do not show much inter-group aggression.

Where, on the other hand, a group of closely related men live together as a social unit, in the same manner as chimpanzees, feuding and raiding between groups is chronic. Among the Yanomamo Indians of Venezuela, for instance, there is almost routine warfare and raiding between villages. In Scottish clans, a McDonald hated a Campbell, and vice versa, long before the massacre of Glencoe gave him an excuse. His descendants, in the suburbs of Glasgow, express the same tribal loyalty to Rangers or Celtic football clubs. After the Second World War it was not logically inevitable that Russians and Americans would come to see each other as enemies and rivals, but it was humanly inevitable. Montagues and Capulets, French and English, Whig and Tory, Airbus and Boeing, Pepsi and Coke, Serb and Muslim, Christian and Saracen – we are irredeemably tribal creatures. The neighbouring or rival group, however defined, is automatically an enemy. Argentinians and Chileans hate each other because there is nobody else nearby to hate.

Indeed so pervasive is the men's habit of them-and-us that human males pursue their quest for status by taking part in battles between groups, whereas chimpanzee males achieve status by battles within the group. Chimpanzee group conflict is not warfare because patrols of rival chimpanzees do not attack each other; they try to find and attack single males instead. They are raids, not battles. Human males have pursued glory in battles with the enemy – from Achilles to Napoleon.[14]

Blue-green allergy

If sporting factions are in effect ersatz battles between rival coalitions of males in a tribal species of ape, the ecstasy and agony of the modern football fan makes a little more sense. The enemy team and its factional supporters are almost as terrifying and provoking a danger to the fan as a group of murderous warriors would be to a Yanomamo. In ancient Roman chariot races in the Circus, the competing chariots were distinguished by the colours of their drivers' liveries. At first there were just two colours, white and red, but these were later complemented, then overtaken, by two more colours, green and light blue. This little device, originally intended to make the chariots more easily distinguished, gave rise to factions of rival supporters within the city. After Caligula, even the emperor often supported one faction or another.

The habit soon spread to Constantinople, where the Hippodrome provided a vast arena for chariot races, and the division of the city into two factions, green and blue, soon followed. This was potent enough, but worse was to come in the sixth century AD. The acid of sporting factions mixed with the alkali of religion and politics, and exploded into internecine fury. The weak but prudent Emperor Anastasius embraced a current heresy and broke with the pope. So his team – the greens – came to be associated with the heresy. At a religious festival at the end of his reign, the greens massacred 3,000 blue supporters, beginning a period of greater than usual violence between the two factions. When Anastasius died he was succeeded by an ambitious soldier, Justin, who was in turn succeeded by his even more ambitious nephew, Justinian, who had married a still more ambitious ex-prostitute named Theodora, who had suffered in her acting days at the hands of the greens. Justinian and Theodora ruthlessly re-imposed religious orthodoxy, while showing great favouritism to the blue cause in sport. Greens therefore embraced the heterodox religion and the political resistance to the new regime. The blues terrorized the city with their persecution of greens and heretics. In 532 a riot broke out in the Hippodrome, which Justinian tried to stop by executing ringleaders on both sides; this only inflamed both factions against him and the so-called Nika riot began. Much of the city was burnt, including St Sophia itself, and a reluctant nephew of Anastasius was 'crowned' emperor by the crowd in the Hippodrome. For five days the city was abandoned to the factions, whose watchword was 'Nika', meaning vanquish. Justinian was on the point of flight from his defended palace, but his redoubtable wife saved the situation. She persuaded the blues to abandon the Hippodrome and then sent two generals to storm it; 30,000 greens died there.[15]

This grandmother of all football riots illustrates that the power of xenophobic group loyalty in the human species is every bit as potent as it is in chimpanzees. And yet we also bring to xenophobia a crucial feature of dolphin society. We form second-order alliances. Indeed, a diagnostic feature of many human societies, including most spectacularly the Western one I inhabit, is that they are 'segmented'. We live in small clans, which come together to form tribes, which come together to form alliances and so on. Clans may bicker and fight, but an external threat causes them to close ranks. There are primate parallels to this, though not among our close relatives, the apes. Hamadryas baboons, for example, live in harems: one male plus several mates and a few juvenile male hangers-on. But the harems come together at night in clans, each clan consisting of two or three closely affiliated harems. And several such clans comprise a troop, which shares a territory. What is unique to bottlenose dolphins and humans, though, is the use of alliances between groups to combat third groups. Just as two coalitions of dolphins may get together to steal a female from a third coalition, so the notion of strategic alliances between human tribes is familiar from all of history: my enemy's enemy is my friend.

Yanomamo Indians frequently conclude treaties between villages that have common enemies. The Molotov–Ribbentrop pact, in which Nazi Germany and Stalinist Russia agreed not to attack each other, so paving the way for Germany to attack Poland and France unhindered, was formally identical to the non-aggression pact that Luit and Yeroen used to undermine Nikkie in Frans de Waal's chimpanzees, or that A and B used against H in Connor's dolphins – except that it was between tribes rather than individuals or trios. These are the sort of instinctively familiar coalition-building tactics of human tribalism that stem from a primate tradition of cooperating in the cause of aggression.

Yet surely we cannot explain foreign policy on the basis of instinct? Not in detail, no. One hopes that our diplomats are concluding treaties that are in our interests, rather than relying on genetic memories of hostility between groups of apes on the savanna. But they take for granted certain things about human nature that we need not, in particular our tribalism. My aim is to convince you to try to step out of your human skin and look back at our species with all its foibles. Then we notice that our politics need not be the way it is, for we need not be tribal at all. If we were truly like dolphins and lived in open societies, there would still be aggression, violence, coalition-building and politics, but the human world would be like a water-colour painting, not a mosaic of human populations. There would not be nationalism, borders, in-groups and out-groups, warfare. These are the consequences of tribal thinking, which itself is the consequence of our evolutionary heritage as coalition-building, troop-living apes. Elephants, curiously, do not live in closed societies either. Females aggregate in groups, but the groups are not competitive, hostile, territorial or fixed in membership: an individual can drift from group to group. It is an intriguing fantasy to imagine ourselves like that. Indeed, female human beings are like that already.

CHAPTER NINE

The Source of War

In which cooperative society proves to have a price: group prejudice

A tribe including many members who, from possessing in high degree the spirit of patriotism, fidelity, obedience, courage and sympathy, were always ready to aid one another, and to sacrifice themselves for the common good, would be victorious over most other tribes; and this would be natural selection.

Charles Darwin, *The Descent of Man*, 1871

In Death Valley, the inhospitable furnace of the eastern California desert, the commonest creature is an ant called the desert seedharvester ant, *Messor pergandei*. These ants live in huge colonies of tens of thousands of individuals, in underground nests that penetrate many metres into the ground. From their burrows they fan out across the desert at dawn and dusk in dense columns to collect seeds, which they store underground. Subsisting on these stores, they can survive many years of drought. Each colony is dominated by a single queen from whose eggs hatches a continuous stream of workers.

Nothing unusual in all this. But when *Messor* ants start a new nest, something rather peculiar happens. Several new queens come together and excavate the new nest jointly. They need not be sisters – indeed they often are unrelated – yet they all happily cooperate in the new joint venture, and all begin breeding together. Suddenly, after some weeks, there is an abrupt change in their behaviour. Civil war breaks out in the colony and the queens turn murderously upon each other. Like the last scene in *Hamlet*, royal murder follows royal murder (although in this case, one queen survives). What has changed?

The explanation of this strange tale is that desert seedharvester ants are fiercely territorial. Each patch of desert is owned exclusively by one colony. Yet, since all the colonies produce new, winged queens at the same time, there is for a brief period a large number of new nests in each vacant territory. A state of open warfare exists between the new colonies, with each sending out raiding parties to steal the eggs and larvae of their neighbours. The stolen brood is brought back and reared at home as 'slaves', adding to the strength of the home colony; the colony from which it was stolen, weakened by the loss of antpower, expires. Eventually, there is only one victorious colony left standing.

This warfare explains the curious, temporary cooperation of the founding queens. The more queens there are in a new nest, the more workers they initially rear. The more workers there are, the greater the chance of defending their own brood and stealing an enemy's. So it pays the joint founder queens to cooperate with each other in order to be in a successful group. Competition between the groups is fiercer than between individual ants. Only when there are no enemy groups left do the individuals assert their selfishness against their co-queens.[1]

To put it in more human terms, an external enemy helps group cohesion. This is a thoroughly familiar idea. In the London Blitz, differences and antagonisms were famously forgotten: German bombs achieved a monolithic loyalty among the British (and vice versa). When the war was over, society fragmented once more and the triumphant greater-goodism of the war years disintegrated into the bickering selfishness of peace, gradually spoiling the promise of socialism. To take a more familiar example, London taxi drivers are notorious, in my experience, for their antagonism towards other drivers and their blatant favouritism to other taxis. The typical cabbie screeches to a halt to let another taxi out into the traffic, whether he knows the driver personally or not, but races to cut off a car that tries the same trick, shaking his fist at it and grumbling to his passenger. The taxi-driver's world is divided into two factions, them and us. So he is nice to 'us' and nasty to 'them'.

The same is true of the rivalry between Apple-Macintosh users and those addicted to IBM PCs. A quite astonishing amount of contempt is heaped upon the latter by the former, who believe their software to be inherently superior. It is largely motivated by tribalism.

The selfish herd

Now an entirely new explanation for human society drifts into view. Maybe cooperation is such a feature of our society not because of close kinship, not because of reciprocity, not because of moral teaching, but because of 'group selection': cooperative groups thrive and selfish ones do not, so cooperative societies have survived at the expense of others. Natural selection has taken place not at the level of the individual but at the level of the band or tribe.

For most anthropologists, this idea is far from new. It has been commonplace in anthropology to argue that a good deal of the cultural baggage of being human is there for a direct purpose – to maintain and enhance the integrity of the band, tribe or society. Anthropologists routinely interpret rituals or practices in terms of their promotion of the good of the group, not the individual. They do so mostly in blithe ignorance of the fact that biologists have thoroughly undermined the whole logic of group selection. It is now an edifice without foundation. Like anthropologists, until the mid 1960s most biologists talked glibly about the evolution by natural selection of traits that were good for the species. But what happens when something is good for the species but bad for the individual? What happens, in other words, in a prisoner's dilemma? We know what happens. The individual's interest comes first. Selfless groups would be perpetually undermined by the selfishness of their individuals.

Consider a rookery. All over Eurasia these cawing, gregarious crows feed on grubs they pick up in pastures and come together in spring to breed in colonies of stick nests built in tall trees. They are immensely sociable. The cawing and calling goes on from dawn to dusk in the rookery, as birds squabble, play and court. So grating is the continuous noise from a flock of rooks that one name for such a flock is a parliament of rooks. In the 1960s a biologist tried to describe rookeries and other aggregations of birds as societies: as wholes greater than the sums of their parts. Rooks, said Vero Wynne-Edwards, gather together to get an idea of their population density and so adjust their breeding effort for the year to ensure that over-population does not result. If they are numerous, every rook lays a small clutch and Malthusian starvation for all is averted. 'The interests of the individual are actually submerged or subordinated to the interests of the community as a whole.' Rook flocks compete with each other, but not individual rooks.[2]

What Wynne-Edwards proposed was probably true, empirically speaking. When population densities are high, clutch size is small. But there is a big difference between this correlation and the cause he inferred. Perhaps, replied an opposing ornithologist, David Lack, when densities are high, food becomes scarce and the birds respond by laying small clutches. Besides, how and why could a rook evolve that put the interest of the population before that of itself? If every rook was practising self-restraint, a mutineer that did not would leave more offspring behind and soon its selfish descendants would outnumber the altruists, so the restraint would disappear.[3]

Lack won the argument. Birds do not restrain their breeding urges for the good of the population. Biologists suddenly realized that very few animals ever put the interests of the group or the species before the individual. Without exception, all those that do are actually putting family first, not group. Ant colonies and mole-rat societies are just big families. So is a wolf pack or a dwarf-mongoose troop. So are the nesting groups of scrub jays and other birds in which the young from the previous year help their parents rear the next brood. Unless they are manipulated by a parasite, or in the case of ants enslaved by another species, the only group that animals ever favour over the individual is the family.

Yet many animals form flocks, schools, shoals, herds and packs that consist of far more than big families. The reason they do so is simple selfishness. Each individual is better off in the herd than outside it for the very pertinent reason that the herd provides alternative targets for predators. There is safety in numbers. The reason herring live in shoals and starlings in flocks is simply to reduce the individual odds of becoming the victim. In aggregate, the effect is negative: the shoaling behaviour of herrings makes them favourite victims of humpback and killer whales, which would never bother chasing single fish. But for the individual, it is always better to hide behind another fish. So shoals and flocks are the product of selfishness, not groupishness.

The reason for rookeries may be a little different: joint defence or the opportunity to follow well-fed individuals back to where they found food. But the principle is the same. Being in the flock is a self-interested not a social act. There is, in short, nothing altruistic about the gregarious or social behaviour of animals unless they are in big families.

'The selfish herd' was William Hamilton's term for it, and he proved the point with an imaginary group of frogs on the rim of a circular pond escaping the attention of a snake in the pond by clustering together. Motivated by nothing more than the desire to get between two other frogs to make them more likely to be eaten, the imaginary frogs ended up in a heap. All aggregations in nature that are not families are selfish herds. Even chimpanzee troops may have come together for this reason: the predators in this case being other members of the same species. The main benefit to chimpanzees of living in large troops is that it provides safety in numbers to reduce the risk of a successful assault on the troop's territory by a rival troop.[4]

When in Rome

To prove that group selection rarely manages to overcome individual selection, consider the fact that the sex ratio of virtually all animals at conception is 50:50. Why? Imagine a species of rabbit that consisted of ten females for every male. Since rabbits are polygamous and male rabbits are not needed to feed or protect their young, the species would thrive, and would breed at almost twice the rate of a normal rabbit. It would soon drive the normal rabbit extinct. So a biased sex ratio would be good for the species.

But now look at it from the point of view of a single female rabbit in this new species. Suppose she had the power to alter the sex ratio of her litter. If she produced only sons, each son would have ten mates and she would have ten times as many grandchildren as her rivals. Soon her son-producing lineage would take over the species, and males would get commoner and commoner, returning the sex ratio to equality. That is why, with rare exceptions which prove the rule, the sex ratio always hovers around 50:50. Any deviation automatically rewards those who would bias the sex ratio back to equality.

Much the same argument applies to human behaviour. Suppose there are a hundred families of Indians living in a South American forest and eating only one kind of food – the pith of a palm tree trunk. This is not implausible, for such food is the staple of some people. Suppose that the palm trees grow slowly and that each family has a rule that only the mature trees may be cut for their pith. In order to prevent starvation, each family obeys a strict policy of two children per married couple, and kills any extra babies, which keeps everybody well fed with mature palm trees. All is well in the slightly totalitarian Eden we have created. The species is looked after, at some cost to individual ambition, and thrives.

Now suppose that for some reason, after many years, one family refuses to do what it is told and rears ten children; they feed them by chopping down immature palm trees. Others do the same and the whole tribe is soon in trouble, but the law-abiding Indians are in just as much trouble as the law-breakers. Indeed, because there are now so many law-breakers, they have a better chance of surviving the ensuing famine than any single family of obedient ones. The suffering is shared or even disproportionately borne by the innocent. The species does not thrive, but the individual does. A potential law-breaker might argue that he would be better off in the long run resisting the temptation, or he might be motivated by community spirit. But can he be sure that others will come to the same conclusion? Can he, in the terms of the prisoner's dilemma, trust them not to defect? For that matter, can he even trust them to trust him not to defect? For if one individual defects, or thinks another might defect, or thinks another might think he will defect, then community spirit collapses and logic leads to a free-for-all.

Remember the bleak lesson of the chromosome, the embryo and

the ant colony. Even in such closely related groups, there is a constant threat of selfish mutiny, suppressed only by elaborate mechanisms to create a lottery among chromosomes, to sequester the germ line in the embryo and to sterilize the worker ants. How much greater is the difficulty of suppressing such mutinies when the individuals are unrelated, free to migrate between groups and capable of reproducing on their own.

It was logic such as this that exposed the fatally weak assumptions behind all group-selection thinking. Only if groups have generation times as short as individuals, only if they are fairly inbred, only if there is relatively little migration between groups and only if the whole group has as high a chance of going extinct as the individuals within it – only when these conditions are met will group selection drown the effect of individual selection. Otherwise selfishness spreads like flu through any species or group that tries to exercise restraint on behalf of the larger group. Individual ambition always gets its way against collective restraint. And there is simply no good example, to this day, of an animal or plant that has been found to practise group selection unless in a clone or closely related family – except in the temporary and passing conditions of new-colony formation in the desert seedharvester ant. Bees risk their lives to defend the hive, not because they wish the hive itself to survive, but because they wish the genes they share with their many sisters in the hive to survive. Their courage is gene-selfish.[5]

In recent years, however, a note of doubt has crept into the certainty with which some biologists trot through this argument. They do not doubt its central truth, but they think they may have found an exception to it, a species in which the unlikely conditions apply that could allow groups of cooperators to have such a large advantage over groups of selfish individuals that they could drive the selfish groups extinct before being infected by them.

That exception is, of course, the human being. What makes human beings different is culture. Because of the human practice of passing on traditions, customs, knowledge and beliefs by direct infection from one person to another, there is a whole new kind of evolution going on in human beings – a competition not between genetically different individuals or groups, but between culturally different individuals or groups. One person may thrive at the expense of another not because he has better genes, but because he knows or believes something of practical value.

Rob Boyd is one of those responsible for the new insight, and as usual it came through game theory. Boyd did his first degree in physics and his second in ecology, bringing mathematical rigour to subjects usually treated more gently by biologists. In the 1980s he teamed up with Peter Richerson, an ecologist expert in the study of plankton, to explore group selection. His interest grew out of a paradox. Prisoner's dilemma games lead to Tit-for-tat. But however you cook the sums, reciprocity produces cooperation only in very small groups of individuals. It is all very well for vampire bats or even chimpanzees, each of which has to keep track of the past generosity of two or three individuals. But human beings, even in tribal societies, interact with scores of other individuals, even hundreds or thousands. Yet human beings still reliably cooperate even in these large, diffuse groups. We trust strangers, tip waiters we will never see again, give blood, obey rules and generally cooperate with people from whom we can rarely expect reciprocal favours. To be a selfish free-rider is such a sensible and successful strategy in a large group of reciprocating cooperators (as the occasional Robert Maxwell demonstrates), that it seems crazy more people do not choose such an option.

So, argued Boyd and Richerson, let us reject reciprocity and look for other explanations for humans to cooperate. Suppose that throughout human history groups of cooperators have been more successful than groups of selfish individuals and have driven the latter extinct with fierce and frequent efficiency. This would have the effect of making it more important to be in a group of selfless individuals than to be self-interested yourself. It would work so long as the differences between the groups persisted, but would be fatally undermined if, through intermarriage, for example, selfish ideas could spread from the selfish groups to the cooperative groups. Even if the creature concerned learns most of its habits culturally, rather than relying on instinct, the same conclusion still applies.

But Boyd and Richerson discovered in their mathematical simulations that there is one kind of cultural learning that makes cooperation more likely: conformism. If children learn not from their parents or by trial and error, but by copying whatever is the commonest tradition or fashion among adult role models, and if adults follow whatever happens to be the commonest pattern of behaviour in the society – if in short we are cultural sheep – then cooperation can persist in very large groups. The result is that the difference between a cooperative group and a selfish group can now persist long enough for the latter to become extinct in competition with the former. Selection between groups can start to matter as much as selection between individuals.[6]

Does conformism sound familiar? I think so. Human beings are terribly easily talked into following the most absurd and dangerous path for no better reason than that everybody else is doing it. In Nazi Germany, virtually everybody suspended their judgement to follow a psychopath. In Maoist China, merely by issuing a series of pronouncements a sadistic leader induced vast numbers of people to do ridiculous things like denounce and attack all school teachers, melt down all cooking pots to make steel, or kill sparrows. These may be extreme examples, but do not comfort yourself with the thought that your own society is immune to fads. Imperial jingoism, McCarthyism, Beatlemania, flared jeans, even the absurdities of political correctness are all telling examples of how easily we can be rendered obedient to the current fashion for no better reason than that it is the current fashion.

Boyd and Richerson then asked themselves why conformism should evolve in the first place. What advantage does it confer on human beings to be so conformist? They suggested that in a species that makes its living in many different ways, it makes a good deal of sense to adopt a tradition of 'When in Rome, do as the Romans do.'

To understand why, consider killer whales. Most animals eat the same kinds of things all over their range. A fox, for instance, seeks out carrion, worms, mice, baby birds and insects – whether it lives in Kansas or Leicestershire. But killer whales are different. Each local population employs a sophisticated strategy to catch its particular prey, but it is a different prey in each case. In the fjords of Norway, killer whales specialize in rounding up shoals of herring with ingenious tricks of cooperative hunting. Off British Columbia, killer whales use a rather different set of tricks to catch salmon. In the sub-Antarctic islands, they feed chiefly on penguins and are very good at taking the penguins by surprise among the kelp. Off the Patagonian coast, they have developed a special skill which youngsters must learn of flinging themselves on to the beach and grabbing sea-lions. The point is that each population does something different, and a killer whale from Norway would starve off Patagonia unless it adopted the local habits.

Human beings have probably always been similarly local in their habits ever since they parted genetic company with the ancestors of chimpanzees about five million years ago. Chimpanzees, after all, show strong local feeding traditions according to what works best where they live, almost as much as killer whales. One group in West Africa cracks nuts with stones; another in the east eats termites caught by 'fishing' with sticks inside termite nests. Conformist transmission of culture is one way of ensuring that you do what works locally – you inherit a disposition to copy your neighbours. A *Homo erectus* woman from the Serengeti who migrated west and joined a band that lived in the edge of the mountain forest would do well to copy her new neighbours in searching for fruit rather than insisting on digging for some kind of tuber not found in her new home.

Yet Boyd notices that imitation is more beneficial when everybody is doing it. Otherwise, if you are the only person imitating, then all you learn is what somebody else has laboriously learnt on their own, not what has been proven to work by hundreds of other people. This creates a problem of how a conformist system could get started in the first place.[7]

In human evolution, then, the habits of local specialization, cultural conformism, fierce antagonism between groups, cooperative group defence and groupishness all went hand in hand. Those groups in which cooperation thrived were the ones which flourished and, bit by bit, the habit of human cooperation sank deep into the human psyche. In the words of Boyd and Richerson, 'Conformist transmission provides at least one theoretically cogent and empirically plausible explanation for why humans differ from all other animals in cooperating, against their own self-interest, with other human beings to whom they are not closely related.'[8]

A million people cannot be wrong, or can they?

In parallel with the evolutionary discovery of conformism, psychologists and economists have discovered it, too. In the 1950s, an American psychologist named Solomon Asch did a series of experiments that tested people's tendency to be intimidated into conforming. The subject entered a room where there were nine chairs in a semi-circle, and was seated next from the end. Eight other people arrived one by one and occupied the other chairs. Unknown to the subject, they were all stooges – accomplices of the experimenter. Asch then showed the group two cards in turn. On the first was a single line; on the second there were three lines of different length. Each person was then asked which of the three lines was the same length as the line they had first seen. This was not a difficult test; the answer was obvious, because the lines were two inches different in length.

But the subject's turn to answer came eighth, after seven others had already given their opinion. And to the subject's astonishment the seven others not only chose a different line, but all agreed on which line. The evidence of his senses conflicted with the shared opinions of seven other people. Which to trust? On twelve out of eighteen occasions the subject chose to follow the crowd and name the wrong line. Asked afterwards if they had been influenced by others' answers, most subjects said no! They not only conformed, they genuinely changed their beliefs.[9]

This clue was picked up by David Hirshleifer, Sushil Bikhchandani and Ivo Welch, who are mathematical economists. They take conformity as read and try to understand why it happens. Why do people follow the local fashion in time and place? Why are skirt lengths, fashionable restaurants, crop varieties, pop singers, news stories, food fashions, exercise fads, environmental scares, runs on banks, psychiatric excuses and all the rest so tyrannically similar at any one time and in any one place? Prozac, satanic child abuse, aerobics, Power Rangers – whence these crazes? Why does the primary-election system of the United States work entirely on the proposition that people will vote for whoever seems to be winning, as judged by the tiny state of New Hampshire? Why are people such sheep?

There are at least five explanations that have been proposed over the years, none of which is very convincing. First, those who do not follow the fashion are punished in some way – which is simply not true. Second, there is an immediate reward for following the fashion, as there is for driving on the correct side of the road. Again, usually false. Third, people simply irrationally prefer to do what others do, as herrings prefer to stay in the shoal. Well, perhaps, but this does not answer the question. Fourth, everybody comes independently to the same conclusion, or fifth, the first people to decide tell the others what to think. None of these explanations begins to make sense for most conformity.

In place of these hypotheses, Hirshleifer and his colleagues propose what they call an informational cascade. Each person who takes a decision – what skirt length to buy, what film to go and see, for instance – can take into account two different sources of information. One is their own independent judgement; the second is what other people have chosen. If others are unanimous in their choice, then the person may ignore his or her own opinion in favour of the herd's. This is not a weak or foolish thing to do. After all, other people's behaviour is a useful source of accumulated information. Why trust your own fallible reasoning powers when you can take the temperature of thousands of people's views? A million customers cannot be wrong about a movie, however crummy the plot sounds.

Moreover, there are some things, such as clothes fashion, where the definition of the right choice is itself the choice that others are making. In choosing a dress, a woman does not just ask, 'Is it nice?' She also asks, 'Is it trendy?' There is an intriguing parallel to our faddishness among certain animals. In the sage grouse, a bird of the American high plains, the males gather in large flocks called leks to compete for the chance to inseminate the females. They dance and strut, bouncing their inflatable chests about with abandon. One or two males, usually the ones holding court near the centre of the lek, are by far the most successful. Ten per cent of the males can perform ninety per cent of the matings. One of the reasons this is so is that the females are great copiers of each other. A male is attractive to females merely because he has other females already surrounding him, as experiments with dummy females easily demonstrate. This faddishness on the part of the females means that the choice of male can be rather arbitrary, but it is none the less vital that they follow the fashion. Any female that breaks ranks and picks a lonely male will, in all probability, have sons who inherit their father's inability to attract a crowd of females. Therefore, popularity in the mating game is its own reward.[10]

Back to human beings. The problem with obeying the information cascade is that the blind can end up leading the blind. If most people are letting their judgements be swayed by others, a million people can be wrong. To argue that a religious idea must be true because other people have been convinced by it for a thousand years is fallacious; most of the other people have been swayed by the fact that their predecessors have been swayed. Indeed, one of the features of human fads that only the Hirshleifer theory can explain is that they are as fragile as they are spectacular. With only the slightest new piece of information, everybody abandons the old fashion for a new one. Our faddishness appears, then, as a rather foolish characteristic, which sends us bouncing from one craze to another at the whim of cascading information.

Yet, in a small band of hunter-gatherers, it might have been a more useful habit to obey the fashion. To a large extent, human society is not a society of individuals, as the society of leopards, or even lions, is – albeit the individual lions are lumped together in groups. Human society is composed of groups, superorganisms. The cohesiveness of groups that conformity achieves is a valuable weapon in a world where groups must act together to compete with other groups. That the decision may be arbitrary is less important than that it is unanimous.[11]

Much the same thought occurred to the computer scientist Herbert Simon. He suggested that our ancestors thrived to the extent that they were socially 'docile', by which he meant receptive to social influence. Remember how we constantly proselytize each other about the virtues of selflessness. If we have been naturally selected to be receptive to this indoctrination, then we are more likely to find ourselves in successful groups by virtue of our altruistic biases. It is cheaper and usually better, says Simon, to do what other people say than to figure out the best way to do something yourself.[12]

Love thy neighbour, but hate everybody else?

If people conform to the traditions of their native groups, then there will be an automatic tendency for each group of people to be culturally different. If one group has a taboo against pork and another a taboo against beef, then conformism will maintain the distinction between the groups. Those who join one group will conform to its taboos. It is therefore easy to get widely divergent practices in competition with each other, each being represented by a group of people. If, further, there is a good chance of groups going extinct in competition with other groups, and if new groups form by the splitting of old groups, rather than the recruitment of people from many groups, then the conditions for group selection look promising.

Do these conditions apply to human beings? Joseph Soltis, a colleague of Boyd and Richerson, set out to test the notion by examining the history of tribal warfare in New Guinea. New Guinea is unusual because most of its tribes first came into contact with Westerners in the last century or this, and were still living in a state undisturbed by Western goods, practices or beliefs when anthropologists first met them. It is therefore hard to argue that the routine practice of tribal warfare was some artefact of Western contact. Most New Guineans lived in a pretty Hobbesian state: violence was an ever-present threat.

Soltis analysed the history of hundreds of conflicts over about fifty years in various parts of the island. In almost all cases it was clear that new groups formed by the splitting of old groups in two, and that tribal warfare frequently caused the extinction of groups. For instance, among the Mae-Enga people of the central Western Highlands, twenty-nine conflicts over fifty years among fourteen clans caused five of those clans to disappear. Not that clan members all died, but they dispersed after defeat and joined other, victorious clans – assimilating rapidly into them. (This incidentally is one reason why genetic group selection does not work – the genes of the defeated individuals survive; indeed, in the case of women captured in war after the sack of an ancient city and taken as wives, the genes of the defeated individuals probably thrived and infiltrated the victorious group. But because the defeated individuals drop their culture and absorb that of the victors, cultural group selection can work.) In all, Soltis calculates that New Guinea clans died out at the rate of between two and thirty per cent every twenty-five years.

This rate of extinction of groups would be swift enough to drive only a very mild form of cultural group selection. The extinction of ill-suited groups by ones with better traditions can explain trends that occur over 500 or 1,000 years, but it cannot explain shorter changes. And most human cultural change is more rapid than that. For instance, the introduction of the sweet potato into the agriculture of New Guinea spread far too rapidly to be accounted for by the selective advantage of those groups that used the potato over those that did not. The potatoes undoubtedly spread by diffusion from tribe to tribe.[13]

There is another difficulty with the group-selection account of human history. As Craig Palmer has argued, human groups are largely mythical. People do undoubtedly think in terms of groups: tribes, clans, societies, nations. But they do not really live in isolated groups. They mingle continuously with those from other groups. Even the clan group beloved of anthropologists is often an abstraction – people know their kin group, but they do not live only with their kin. In patrilineal societies, people live with their father's relatives, but they still surely absorb some culture from their mothers. Human groups are fluid and impermanent. People do not live in groups, says Palmer, they merely perceive the world in terms of groups, ruthlessly categorizing people as us or them. Yet this is a double-edged discovery. That we see the world in terms of groups – however falsely – still tells us something about the human mindset, and it is inside the skull that evolution leaves many of its social marks.[14]

There is one final nail in the coffin of human group selection theory. To argue that human beings are conformist and therefore share their fate with that of the group is facile. Most of the examples I have discussed are cases where individuals are cooperating to further their self-interest. That is not group selection: it is individual selection mediated by groupishness. Group selection occurs when individuals cooperate against their own self-interest but in the interest of the group – they show self-restraint in breeding, for instance. All we have identified in human beings is a powerful tendency to be groupish in the pursuit of individual goals, not evidence of putting groups before individuals. A mind that has been selected to gain the advantages of living in groups (conformism is an instance) is not the same as a mind that has evolved by group selection. Groupishness can enhance individual selection – but that is not group selection.

The problem arises, according to John Hartung, because we are so instinctively groupish that we prefer to pretend – and perhaps even believe – that we are group-selected. In other words, people claim they are putting the interests of the group first and not their own interests, the better to disguise the fact that they only go along with the group when it suits them. Pointing this out to them makes you unpopular, as every Hobbesian since Hobbes has discovered.

The fact that people form emotional attachments to groups, even arbitrary ones, such as randomly selected school sports teams, does not prove group selection, but the reverse. It proves that people have a very sensitive awareness of where their individual interests lie – with which group. We are an extremely groupish species, but not a group-selected one. We are designed not to sacrifice ourselves for the group but to exploit the group for ourselves.[15]

Take your partners

Open virtually any illustrated book about anthropology and you will be confronted with pictures of dance, magic, ritual and religion. You may search in vain for details of what a particular tribe does at mealtimes or how men court women or how children are brought up. This is no accident, for traditions of eating, courting and child-rearing differ rather little between tribes and societies across the world. But creation myths, ways of body painting, head-dress design, magical incantations and dance patterns all show marked cultural peculiarities. They are what distinguish one people from another and they are far from incidental to the lives of people; huge amounts of time, effort and prestige are invested in these things. They are what people live for. Yet all people have them; it would be as odd to find a tribe in New Guinea to whom the words dance, myth or ceremony (suitably translated) meant nothing at all as it would be to find one that did not know the meaning of hunger, love or family. Ritual is universal; but its details are particular.

I am about to argue that one way to understand ritual is as a means of reinforcing cultural conformity in a species dominated by groupishness and competition between groups. Humankind, I suggest, has always fragmented into hostile and competitive tribes, and those that found a way of drumming cultural conformity into the skulls of their members tended to do better than those that did not.

The anthropologist Lyle Steadman argues that ritual is about more than demonstrating the acceptance of tradition; it is also specifically about the encouragement of cooperation and sacrifice. By taking part in a dance, a religious ceremony or an office party, you are emphasizing your willingness to cooperate with other people. A sportsman sings the national anthem before taking the field; a parent submits to 'trick-or-treat' humiliation at Hallowe'en; a homeowner opens his doors to carol singers at Christmas; a senior doctor laughs through clenched teeth at jokes about him in a play by medical students at the end of term; a church-goer sings a hymn in unison with his neighbours during a service; a football crowd does the Mexican wave: in every case, the far-from-incidental message is there for all to see. It says: we are part of the same team; we are on the same side; we are all one.[16]

Nothing so clearly shows this as dance, which is no more and no less than people moving in unison with each other as aided by the rhythm of the music. The historian William McNeill argues that the otherwise inexplicable human fascination with dance must have something to do with asserting cooperative spirit, binding people together emotionally and rehearsing the identity of the group. Dancing in preliterate societies in Africa, Asia and South America has little to do with courtship or sexual display. It is an act of ritual that emphasizes team spirit. A South African crowd making a political demonstration and jogging in musical rhythm is thus much closer to the roots and purpose of dance than a ballroom of Viennese waltzing the night away.[17]

Much the same argument has been made for the origin of music, for example by the philosopher Anthony Storr. Music is arousing, and emotionally moving, in predictable and universal ways, which is why it accompanies films to enhance the power of the scene. 'Rhythm and harmony find their way into the inward places of the soul,' thought Socrates. St Augustine agreed, adding that it was a grievous sin to find the singing in church more moving than the truth it conveyed. The great conductor Herbert von Karajan was once wired up during concerts to record his pulse rate. It varied according to the mood of the music rather than the energy of the conducting; when he piloted and landed a jet aircraft, his pulse rate varied less than when he conducted.

So music stirs the emotions. The evolutionary benefit of letting the emotions be stirred by music may well be to synchronize and harmonize the emotional mood of a group of individuals at a time when they are called upon to act in the interests of the group, the better to further their own interests. The Pythagorean philosophers called music the reconciliation of the warring elements. It is no accident perhaps that music is intimately associated with displays of group loyalty – even more than dance. Hymns, football chants, national anthems, military marches: music and song were probably associated with group-defining rituals long before they served other functions. There may even be an animal that has similar reactions to rhythm and melody. Gelada monkeys, or bleeding heart monkeys, live in very large troops on the high grasslands of Ethiopian mountains, where they eat a diet of grass. They respond with cohesion and collective purpose to melodic singing by members of the group. In human beings, in a similar fashion, 'A culturally agreed upon pattern of rhythm and melody, i.e. a song, that is sung together, provides a shared form of emotion that, at least during the course of the song, carries along the participants so that they experience their bodies responding emotionally in very similar ways.'[18]

As for religion itself, the universalism of the modern Christian message has tended to obscure an obvious fact about religious teaching – that it has almost always emphasized the difference between the in-group and the out-group: us versus them; Israelite and Philistine; Jew and Gentile; saved and damned; believer and heathen; Arian and Athanasian; Catholic and Orthodox; Protestant and Catholic; Hindu and Muslim; Sunni and Shia. Religion teaches its adherents that they are a chosen race and their nearest rivals are benighted fools or even subhumans. There is nothing especially surprising in this, given the origins of most religions as beleaguered cults in tribally divided, violent societies. Edward Gibbon noticed that a vital part of Roman military success was religion: 'The attachment of the Roman troops to their standards was inspired by the united influences of religion and of honour. The golden eagle, which glittered in the front of the legion, was the object of their fondest devotion; nor was it esteemed less impious than it was ignominious, to abandon that sacred ensign in the hour of danger.'[19]

John Hartung, an anthropologist who pursues his training as a historian in his spare time, has taken the much loved Judaeo-Christian phrase 'Love thy neighbour as thyself' and subjected it to searching scrutiny. It was devised, according to the biblical account in the Torah (Old Testament), at a time when the Israelites were in the desert, rent by dissension in the ranks and devastated by internecine violence. Three thousand people had died in a recent episode. Moses, anxious to maintain amity within the tribe, came up with the pithy aphorism about loving neighbours, but the context of his remark is clear. It refers directly to 'the children of thy people'. It does not profess general benevolence. 'A parochial perspective characterizes most religions,' says Hartung, 'because most religions were developed by groups whose survival depended upon competition with other groups. Such religions, and the in-group morality they foster, tend to outlive the competition that spawned them.'

Hartung does not stop there. The ten commandments, he reveals, apply to Israelites but not heathen people, as reaffirmed throughout the Talmud, by later scholars such as Maimonides and repeatedly by the kings and prophets of the Torah. Modern translations, by footnotes and judicious editing or mistranslation, usually blur this point. But genocide was as central a part of God's instructions as morality. When Joshua killed twelve thousand heathen in a day and gave thanks to the Lord afterwards by carving the ten commandments in stone, including the phrase 'Thou shalt not kill', he was not being hypocritical. Like all good group-selectionists, the Jewish God was as severe towards the out-group as he was moral to the in-group.

This is not to pick on the Jews. No less an authority than Margaret Mead asserted that the injunction against murdering human beings is universally interpreted to define human beings as members of one's own tribe. Members of other tribes are subhuman. As Richard Alexander has put it, 'the rules of morality and law alike seem not to be designed explicitly to allow people to live in harmony within societies but to enable societies to be sufficiently united to deter their enemies.'[20]

Christianity, it is true, teaches love to all people, not just fellow Christians. This seems to be largely an invention of St Paul's, since Jesus frequently discriminated in the gospels between Jews and Gentiles, and made clear that his message was for Jews. St Paul, living in exile among the Gentiles, started the idea of converting rather than exterminating the heathen. But the practice, rather than the preaching, of Christianity has been less inclusive. The Crusades, the Inquisitions, the Thirty Years War and the sectarian strife that still afflicts communities like Northern Ireland and Bosnia, testify to a continuing tendency for Christians to love only those neighbours who share their beliefs. Christianity has not notably diminished ethnic and national conflict; if anything, it seems to have inflamed it.

This is not to single out religion as the cause or source of tribal conflict. After all, as Sir Arthur Keith pointed out, Hitler perfected the double standard of in-group morality and out-group ferocity by calling his movement national socialism. Socialism stood for communitarianism within the tribe, nationalism for its vicious exterior. He needed no religious spur. But given that humankind has an instinct towards tribalism that millions of years of groupishness have fostered, religions have thrived to the extent that they stressed the community of the converted and the evil of the heathen. Hartung ends his essay on a bleak note, doubting that universal morality can be taught by religions steeped in such traditions, or that it can even be attained unless a war with another world unifies the whole planet.[21]

If human beings are nice to each other only because of an inherent xenophobia learnt during millennia of lethal inter-group violence, then there is not much comfort here for moralists. Nor is there much encouragement for those who would urge us to do things for the human race, or Gaia, the whole planet. As George Williams has pointed out, preferring the morality of group selection to the ruthlessness of individual struggle is to prefer genocide over murder. Ants and termites have not, as Kropotkin put it, renounced the Hobbesian war; they merely carry it on between armies rather than individuals. Naked mole rats, so harmoniously sociable within the colony, are notoriously aggressive to mole rats from other colonies. Flocks of starlings, in contrast, have no grudge against other flocks. It is a rule of evolution to which we are far from immune that the more cooperative societies are, the more violent the battles between them. We may be among the most collaborative social creatures on the planet, but we are also the most belligerent.

This is the dark side of groupishness in human beings. But there is a bright side, too. Its name is trade.

CHAPTER TEN

The Gains from Trade

In which exchange makes two plus two equal five

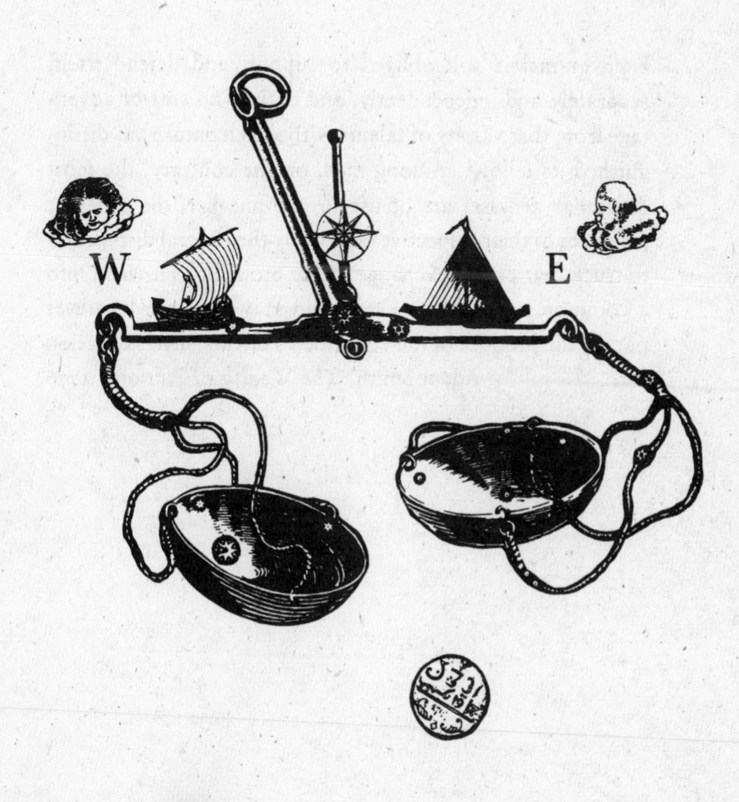

Each animal is still obliged to support and defend itself, separately and independently, and derives no sort of advantage from that variety of talents with which nature has distinguished its fellows. Among men, on the contrary, the most dissimilar geniuses are of use to one another; the different produces of their respective talents, by the general disposition to truck, barter and exchange, being brought, as it were, into a common stock, where every man may purchase whatever part of the produce of the other men's talents he has occasion for.

Adam Smith, *The Wealth of Nations*, 1776

The Yir Yoront Aboriginals live at the mouth of the Coleman River on the York Peninsula in northern Australia. Until recently, they were literally living in the Stone Age. They possessed no items made of metal. They were also true hunter-gatherers, who lived by hunting game, catching fish and gathering vegetable food in the forest. They had no agricultural crops and only one domesticated animal, the dog. They lived under no system of government and answered to nothing that might be called the law. They had, therefore, none of the great inventions to which we attribute the origin of our civilization: no iron, no state, no farming, no judicial system, no writing, no science.

Yet they had one thing we would consider modern, a thing we usually assume cannot be carried out without a state, without a judicial system and without writing. That thing was a sophisticated system of trade.

The Yir Yoront used polished stone axes, carefully hafted into wooden handles. The axes were highly valued, and in almost constant use. Women used them to gather wood for the camp fires, to build and mend their wet-season huts and to dig for roots or cut trees for fruit and fibre. Men took them hunting and fishing, or to hack out wild honey from crevices in trees, and for making secret objects for use in ceremonies. The axes belonged to the men, and were borrowed by the women.

Yet the Yir Yoront lived upon a flat alluvial coast. The nearest quarries from which suitable stone for making axes could be mined were four hundred miles inland to the south. Between the Yir Yoront and the quarries lay many other tribes. Conceivably, the Yir Yoront could have walked south once every few years to gather new stones for making axes, but it would have been a great risk and waste of time for them. Fortunately, there was no need. Plenty of stone axes reached them from the tribes that lived around the quarries, because there was a long line of trading partners who passed them on in exchange for other goods that went south through the same hands. Indeed, the Yir Yoront were not the end of the chain. Their neighbours to the north relied on them to pass on stone axes. Meanwhile, spears tipped with the barbs of stingrays went the other way: southwards.

The trading was man-to-man, each individual man having a partner in the neighbouring tribe. It worked not because of some overall plan by the Yir Yoront to produce stingray barbs and trade them for axes, but because of a simple question of price. A Yir Yoront could buy one stone axe head from his southern neighbour for a dozen barbed spears. He could sell an axe head to his northern neighbours for more than a dozen spears. He could therefore make a profit on the deal. So he tended to pass the axes north. As his spears worked their way south, their value rose relative to that of the stone axe heads. One hundred and fifty miles inland, one spear was worth one axe head. By the time it reached the quarry, it was probably (nobody recorded the truth) worth a dozen axe heads. Most of the people through whose hands the items passed manufactured neither axe heads nor spears. But it is not hard to see that they could make a handsome profit (i.e. keep some axe heads and some spears) simply by acting as middlemen. They had discovered arbitrage: buying something where it is cheap and selling it where it is dear.

By the end of the nineteenth century, apart from an occasional bloody skirmish with white settlers, the Yir Yoront were still largely untouched by the modern world. But already they had steel axes which had begun to work their way north from camps where they were distributed by missionaries. Steel axes were so superior to stone ones that they cost far more. Desperate to obtain them, the Yir Yoront were reduced to drastic measures in their attempt to raise sufficient funds. The tribal gatherings during the dry season, when in the past men had obtained a year's supply of stone axes from their partners, now became rather less enjoyable. To get a single steel axe, a Yir Yoront man might prostitute his wife to a total stranger.[1]

Trade wars

The trading system of the Yir Yoront was by no means unusual for Stone Age man. But it illustrates two things of great importance. First, trade is the expression of the division of labour. Catching stingrays was easy for the Yir Yoront; quarrying stones was easy for the tribes near the quarries. If each tribe did what it was good at and exchanged the result, both were better off. And so were the middlemen in between. In just the same way, a worker ant and a queen are both better off for the fact that each is specialized to its job, and your body works better because the stomach does what it is good at and pools the resulting rewards with other parts of the body. Life, as we have said before, is not a zero-sum game – that is, there does not have to be a loser for every winner.

The second lesson of the Yir Yoront story is that there is nothing modern about commerce. For all the protestations of Karl Marx and Max Weber, the simple idea of gains from trade lies at the heart of both the modern and the ancient economy, not the power of capital. Prosperity is the division of labour by trade; there is nothing else to it. Thousands of years before Adam Smith and David Ricardo were born, human beings had discovered this truth and were exploiting it. The Yir Yoront were, as Rousseau and Hobbes would both agree, in a 'state of nature'. Yet no despotic monarch had imposed a social contract on them, as Hobbes thought necessary; and nor did they live in asocial bliss, as Rousseau fantasized. On the contrary, trade, specialization, the division of labour and sophisticated systems of barter exchange were already part of a hunter-gathering life. Indeed, they had probably been so for many hundreds of thousands of years. Perhaps even millions. It is possible that *Homo erectus* was mining stone tools at specialized quarries, presumably for export, 1.4 million years ago.

Man the hunter-gatherer, man the savanna primate, man the social monogamist – and man the exchanger. Exchange for mutual benefit has been part of the human condition for at least as long as *Homo sapiens* has been a species. It is not a modern invention.

Yet you may search the anthropological literature in vain for an admission that trade is both ancient and common in pre-industrial people. There is a simple reason for this, which the Yir Yoront neatly illustrate. By the time an anthropologist arrives on the scene, the trading patterns have long since been disrupted by the advent of Western goods. The Yir Yoront got their first steel axes before they regularly saw white men. So trade has been consistently under-estimated by students of how our ancestors lived in hunter-gathering times.[2]

Trade is the beneficent side of human groupishness. I have argued that human beings, along with chimpanzees, are unusual in their addiction to group territoriality and inter-group conflict. We segregate into territorial groups and the shared fate that we enjoy with other members of the group drives us into a mixture of xenophobia and cultural conformity, an instinctive subservience to the larger whole that partly explains our collaborative nature.

But this segregation into groups also allows trade between specialized groups. Chimpanzee troops are closed: there is no interchange between them except through violence and emigration. Human groups are not and never have been so closed; they are permeable.[3] People from different bands meet to exchange goods, information and food, as well as to fight. The goods they tend to exchange are those that are scarce or unpredictable in supply. But in some cases, they even seem to invent a need for exchange in order to stimulate trade. The most illuminating example of this is the case of the Yanomamo, studied by Napoleon Chagnon in the Venezuelan rain forest.

Chagnon argues that the Yanomamo live in a state of chronic warfare between villages. Violent death is common among men, abduction common among women. But it is not like chimpanzee warfare, a Hobbesian war of each group against all groups. It is far subtler than that. The key to success for a Yanomamo village is alliance with another village. A complex network of variously intimate ententes binds together different villages into competing alliances. Just as chimpanzee and dolphin individuals succeed by building alliances between individuals, so human groups succeed by building alliances between groups.

The glue of such alliances is trade. Chagnon believes that Yanomamo villages deliberately engineer a division of labour between them in order to provide an excuse for trade, which then seals the political alliance between them.

> Each village has one or more special products that it provides to its allies. These include items such as dogs, hallucinogenic drugs (both cultivated and collected), arrow points, arrow shafts, bows, cotton yarn, cotton and vine hammocks, baskets of several varieties, clay pots, and, in the case of contacted villages, steel tools, fishhooks, fishline, and aluminium pots.[4]

This is not because each village has better access to particular raw materials. Every village could in principle supply its own wants. But the people deliberately choose not to, because – though Chagnon thinks it is not necessarily a conscious motive – it helps stimulate trade and hence alliances. He gives the example of a village that relied on an allied village for its clay pots, and whose people claimed they could not make, or had forgotten how to make, such pots. However, when they fell out with their allies, they quickly remembered the skill of pot-making for themselves. Yanomamo villages largely trade artefacts, not food. I suspect this is a universal feature of early trade – that it relied on a technological division of labour, not an ecological one.

The Yanomamo people have been gardening and hunting in the Amazon for a relatively brief time, probably less than ten thousand years, whereas the Aboriginals have been hunting and gathering in Australia for perhaps six times as long. But there are remarkable similarities between the patterns of trade in both Stone Age peoples, including especially the association between trade and reciprocal feasting. Chagnon believes that it is the feasting that is, so to speak, the aim, and trade that is the excuse, because from the feasts comes the friendship that cements the alliance that is valuable in warfare.

But whether trade is the means or the end hardly matters. The same lesson applies: trade is the precursor of politics, not the consequence.

The merchant law

This is a startling discovery. If trade precedes law, then a whole house of philosophical cards collapses. Jeremy Bentham said: 'Before the law there was no property: take away the law and all property ceases.' Even the most rabid free-trader has been wont to argue that government must hold the ring, enforcing contracts between merchants in an industrial economy. Without recourse to the law and protection from government, commerce is fragile and will disappear.

Yet this is back to front. Government, law, justice and politics are not only far more recently developed than trade, but they follow where trade leads. Indeed, just as this is true for hunter-gatherers, so it now appears to have been true for medieval merchants as well. Modern commercial law was invented and enforced not by governments, but by merchants themselves. Only later did governments try to take it over, and with mostly disastrous results.

Go back to eleventh-century Europe. Agricultural productivity had improved thanks to various innovations; the result was that surplus labour had left the land and moved to the towns to work on the manufacture of goods other than food; exchanging these goods made by the artisans for the food grown by the farmers benefited both, and gave a further spin to the engine of prosperity. For the first time the volume of trade created a new class of prosperous and professional merchants. As the economic expansion continued some of these merchants began to look abroad for opportunities to exploit comparative advantages between countries. But a merchant in a foreign country had no recourse to his sovereign if cheated and no confidence that the same standards applied there as at home. So merchants began to get together and formulate the rules of the game. The *lex mercatoria* was born. It had no recognition from the state. It was voluntarily produced, voluntarily adjudicated and voluntarily enforced. It was like the customs of a club.

It evolved. Good customs that worked, and good ways of settling disputes, drove out bad by natural selection. By the mid twelfth century, merchants travelling abroad had substantial protection in disputes with local merchants under the merchant law. The only and final sanction against a transgressor was ostracism, but as we have seen ostracism can be a powerful force. A merchant with a reputation as a cheat could not carry on his trade. Merchants formed their own courts, which were more efficient and uniform than the royal and state courts. A set of standardized customs about how bills should be settled, interest paid and disputes resolved obtained all across the continent – and all without the slightest direction from above. Monotony without monopoly.

By the twelfth century commercial middlemen were using the new concept of credit. This was a great improvement on barter, and on money, which had lost the uniformity and fungibility it had had in Roman times. Bankers had begun to emerge, together with mortgages, contracts, promissory notes and bills of exchange. All these were governed by merchant law, not government law. Governments had not even woken up to what was going on. An entirely private, voluntary and informal system of exchange had developed.

In a flurry, government then began to act. It enacted into national laws these merchant customs, allowed appeals to royal courts – and of course took the credit. Henry II of England was not a great law giver; he was a great law nationalizer. Merchant courts immediately lost their power, because of the threat of appeal to higher, royal courts, and the adaptability of the system was lost. For the law to change now required more than the evolution of a new custom to replace the old; it required acts of kings and parliaments. The growing cost and gradual congestion of the official courts soon deprived the system of its speed and frugality.

In latter times, commercial arbitration, born in Liverpool among merchants with claims based on the disruption of the American civil war to the cotton trade, began to bypass the congested courts. Private justice, rent-a-judge style, has been a growth industry in America for some years. Only the restrictive practices of lawyers prevent the civil law gradually privatizing itself again. The lesson, for scientists, though, is crystal clear. Markets, exchanges and rules can develop before government or any other monopolist has defined their rules. They define their own rules, because they have been part of human nature for many millions of years.[5]

Silver and gold

From the Yir Yoront exchanging stingray barbs for stone axes to George Soros speculating against the pound on the foreign-exchange markets is no step at all. They are both engaged in simple arbitrage: buying cheap and selling dear. The fact that one is exchanging useful goods while the other is swapping electronic messages that could in theory be exchanged for inflammable paper currency of no possible practical use is just a distinction without a difference. Money is a proxy for goods.

Half way between the Yir Yoront and George Soros and neatly bridging the gap stands a corrupt fifteenth-century French official, named Jacques Coeur. Coeur was the chief *argentier* to King Charles VII of France, responsible for minting his silver coins. This placed him in an eminently profitable position, of which he took full advantage. Because he was finally had up for his corruption in 1453, we have a glimpse of his business in the records of his trial. His principal route to riches was to ship galleys out of Marseille loaded to the gunwales with silver coins, sell the coins in Syria and buy gold coins with them to bring back to France. One of his ships carried nearly 10,000 silver marcs.[6]

Why? Coeur himself explained: '*Il a prouffit à porter argent blanc en Suyrie, car quand il vault 6 escus par deça il en vault 7 par dela.*' In other words, the same quantity of silver would buy fourteen per cent more gold in Syria than in France. This more than covered the cost and risk of shipping it across the Mediterranean, especially since Coeur could alloy a little copper in with his silver to make it go further, while imprinting the fleur-de-lis on it falsely to certify its purity.

The reasons for this imbalance in price are as fascinating as they are revealing. Step back five hundred years from Jacques Coeur to the end of the first millennium AD. At that time silver coins had virtually disappeared from the Arab world and gold coins had virtually disappeared from Christendom. This reflected the abundance of good mines and rulers' ability to mint good-quality coins. Reflecting Europe's demand for silver, and the East's for gold, gold tended to be worth more, in terms of silver, in the Muslim world than in Christendom.

So things might have continued, but for the Crusades. The crusaders took with them as much gold as they could, but mostly they took silver coins to pay for their exploits. Once established in the Levant they began to mint silver coins. Muslim traders with whom they dealt soon came to possess many of these coins and to use them in their own commerce; likewise, the crusaders came to use gold coins captured or acquired from the Arabs.

The crusaders also minted their own gold coins, usually to a lower standard, sometimes using captured dies, and this began to debase the gold currency of the Arab nations as Gresham's Law went to work. No matter: so much silver had now reached the Arab kingdoms from the crusaders that it became possible to reintroduce their own silver coinage for the first time in more than a century. This, ironically, increased the demand for silver, reversing the differential between the gold price of silver in Europe and the gold price of silver in the East.

Once this happened, a very profitable opportunity arose for entrepreneurs. They minted counterfeit Arab silver coins either in Christian enclaves such as Acre, or back in Europe itself, shipped them East and sold them for gold. These coins, called *millares*, carried the legend 'There is no God but Allah; Mohammed is his apostle; the Mahdi is our Imam', yet they were minted by French and Italian counts, dukes and even bishops in places like Arles, Marseille and Genoa. The pious St Louis, king of France, appalled by this blasphemy, prodded a reluctant pope, Innocent IV, into banning the practice during the 1260s, but it continued underground.

During the thirteenth century approximately three billion *millares* were struck in the Christian world for use in the Arab world, or four thousand tonnes of silver. This equated to twenty-five years' worth of peak production from Europe's silver mines. Whole mines in Serbia, Bosnia, Sardinia and Bohemia devoted their output purely to the *millares* trade. Little wonder that the silver coins of Europe came under increasing strain. Since the most profitable thing to do with a silver coin in France was to take it south for reminting as a *millare*, rulers found it harder and harder to sustain a supply of good coins in their own realms; they gradually debased their currencies instead.

How on earth could the Arabs pay for all this silver? Easy: in gold. To the gold mines of Arabia and central Asia was now added the output of camel trains across the Sahara, bringing gold from Ghana. So much gold came this way that at times in Egypt gold reached parity with silver and even salt. Put yourself now in the position of an Italian ruler. Faced with a desperate shortage of silver, but with vast quantities of gold swilling about among your subject merchants, who have acquired it in exchange for silver *millares*, the sensible thing for you to do is to start minting gold coins instead. Venice and Genoa did so in 1252 and within a century most of Europe had followed suit. But this only made the trade more profitable by increasing the demand for gold. In 1339, the year most German rulers began to mint gold coin, one gram of gold was worth twenty-one of silver. In Syria or Egypt it was worth ten or twelve at the very most.

These extraordinary tides of exchange, known as the bimetallic flows, seem meaningless in the extreme. Money is money whatever it is made from. If, as I have argued, trade is an ancient human habit that enables us to profit from the division of labour across vast distances, what is the point of swapping gold for silver? You cannot eat either. If by some supernatural event, silver had never existed, so that there was only one incorruptible metal available, all the waste of energy and enterprise in the bimetallic flows would have been avoided and merchants could have concentrated on arbitraging goods instead: silk for wheat, say. The bimetallic flows were the medieval equivalent of the currency markets.[7]

There is a genuine distinction between the Yir Yoront and the computer trade on the one hand and Messrs Soros and Coeur on the other, and it is this. Whereas the Yir Yoront's trade makes both sides better off, and so did the shipment to me from Japan of the computer on which I write this sentence, the same cannot be said of speculation on the currency markets. Mr Soros's profit was a straight transfer from the idiotic government that thought it could fix the exchange rate of its currency. Mr Coeur's was a straight transfer from the French economy, whose silver he effectively stole. Trade is a non-zero-sum procedure because of the division of labour; without a division of labour, trade is zero-sum.

Only compare

There is, according to a distinguished modern economist, only one proposition in the whole of social science that is both true and non-trivial.[8] It is David Ricardo's Law of Comparative Advantage. It is highly counter-intuitive, because it leads to the conclusion that one country can have a comparative advantage in some product even if it is less efficient at making it than its trading partner.

Suppose there are only two commodities being traded: spears and axes. One tribe, called – for the sake of argument – Japan, is good at making spears and very good at making axes; the other, called Britain, is bad at making spears and very bad at making axes. Superficially, it seems to make sense for the first lot to make their own spears and axes and not indulge in trade at all.

But hold on. A spear is worth a certain number of axes. Let us say one spear is worth one axe. So every time the first tribe makes a spear, it is making something it could buy from the other tribe by making an axe. Since it takes this tribe less time to make an axe than a spear, it would be sensible to make an extra axe, instead of a spear, and swap it for a spear made by the second tribe. The second tribe reasons likewise. Every time it makes an axe, it could have achieved the same result by making a spear more quickly and swapping it with the first tribe for an axe. So if the first tribe specializes in axes and the second in spears, both tribes are better off than if each tries to be self-sufficient. This is true, despite the fact that the first tribe is better at making spears than the second tribe.

This was Ricardo's insight. David Ricardo was a man of some success. Born in London in 1772, the son of a Dutch merchant banker, he went to work for his father at the age of fourteen, later fell in love with a Quaker girl and converted from Judaism to marry her. At twenty-two he set up in business on his own and began to speculate on the stock market with £800. In four years he was wealthy and in twenty he had made a fortune variously estimated at between £500,000 and £1,600,000. His trick, apparently, was to profit from the fact that other investors overreact to news: so he bought when the news was good and sold when it was bad, profiting from the fact that others would do likewise. In 1815 he made another fortune by buying heavily in the government securities market on the presumption that Wellington would win the Battle of Waterloo.[9]

When Ricardo entered parliament as a radical in 1819, he rapidly developed a reputation as the best economist in the House of Commons, where he championed the cause of free trade, but with little success. He did not live to see the Corn Laws repealed in 1846.[10]

Ricardo's law of comparative advantage is such a surprising idea that to this day politicians would be laughed at if they were to assert it. Yet it is trivially easy to demonstrate that it must be true. Winston Churchill was a decent bricklayer – better than many bricklayers (it's true, he was) – but it still paid him to buy most of his brick-laying services from a professional because he was an even better politician. The implications for trade policy are obvious. Even if Japan is better at manufacturing everything conceivable than Britain is, there will still be things that it pays Japan to buy from Britain, because Japan can get more of them by swapping the things it is best at than by making them itself.[11]

I may seem to have rather belaboured the point. If the law of comparative advantage has been known since 1817, why, you may ask, is he announcing it as if it were a newly minted insight? My purpose, though, is not to rehash the argument against protectionism and in favour of free trade that is the usual sequel to a discussion of David Ricardo, but to emphasize the relentless virtues of specialization at the level of the group as well as the individual. If, as I have argued, trade has been going on for hundreds of thousands of years, the reason lies in David Ricardo's comparative advantage. Most anthropological discussions make an assumption of autarky. They depict hunter-gatherer man, squatting on the savanna, entirely self-sufficient in all his needs. They recognize, perhaps, a division of labour between husband and wife, perhaps even between good hunters and good honey-finders, but not between one band and another. I wonder if this is unfair. How do we not know that the savanna was home to many different kinds of bands of human beings? Near the shores of the lake where Olduvai Gorge now stands, there might have been fishermen, engaged in a lively trade swapping reed baskets for bone hooks from the big-game hunters further inland, who in turn traded hides for stone with the men who lived in the forests to the west, and so on, all across the continent.

There are good theoretical reasons for believing that a division of labour between groups is even more productive than one within a group. Sharing between individuals reduces the risk of shortage each individual faces. But resources are much more likely to have been short for one whole band at the same time than for distant bands or bands specializing in different activities. A drought may have harmed the hunting but made fishing easier. The old Adam Smith arguments for divisions of labour apply with equal force between groups and within groups.[12]

By 200,000 years ago stone tools were travelling long distances from their quarries. By about 60,000 years ago, early in the so-called Upper Paleolithic revolution, when modern human beings, equipped with rapidly changing technologies, spilled out of Africa and replaced more antique kinds in Europe and Asia, other goods began to appear regularly much more than a day's walk from where they were manufactured. By 30,000 years ago in Europe, pierced sea shells, to be used as beads, were travelling 400 miles inland or more to turn up in grave goods and other such places. Perhaps it is no coincidence that for the first time there is evidence of specialization between different settlements at about the same period. Where the

Neanderthals all lived in much the same fashion, their replacements began to show great local variation in their stone technologies and their styles of art. It seems to represent the beginning of Ricardo's comparative advantage.[13]

Even if I am wrong, even if trade between groups came much later, at the brink of recorded history, its invention represents one of the very few moments in evolution when *Homo sapiens* stumbled on some competitive ecological advantage over other species that was truly unique. There simply is no other animal that exploits the law of comparative advantage between groups. Within groups, as we have seen, the division of labour is beautifully exploited by the ants, the mole rats, the Huia birds. But not between groups.

David Ricardo explained a trick that our ancestors had invented many, many years before. The law of comparative advantage is one of the ecological aces that our species holds.

CHAPTER ELEVEN

Ecology as Religion

In which living in harmony with nature proves harder than expected

The good shepherd giveth his life for his sheep. But he that is an hireling, and not the shepherd, whose own the sheep are not, seeth the wolf coming, and leaveth the sheep, and fleeth: and the wolf catcheth them, and scattereth the sheep. The hireling fleeth, because he is an hireling, and careth not for the sheep. The Gospel according to St John 10.11–13

Chief Seattle, leader of the Duwamish Indians, delivered a famous speech to the governor of Washington territory in 1854. The governor had offered to buy the chief's land on behalf of Franklin Pierce, president of the United States. Seattle replied in a long and shaming speech that is now among the most widely quoted texts in all environmental literature. It presages almost every thread in the philosophy of the modern conservation movement. The speech exists in various slightly different versions, one of the most moving being that which Albert Gore quoted in his book *Earth in the Balance*:

How can you buy or sell the sky? The land? The idea is strange to us . . . Every part of this earth is sacred to my people. Every shining pine needle, every sandy shore, every mist in the dark woods, every meadow, every humming insect. All are holy in the memory and experience of my people . . . Will you teach your children what we have taught our children? That the earth is our mother? What befalls the earth befalls all the sons of earth. This we know: the earth does not belong to man, man belongs to the earth. All things are connected like the blood that unites us all. Man does not weave the web of life, he is merely a strand in it. Whatever he does to the web, he does to himself.[1]

To Gore this shows 'the rich tapestry of ideas about our relationship to the earth' contained in native American religions. For Gore, as for so many people today, respect for the earth is not just good sense, but a sort of moral virtue. To question it is to sin: 'We each need to assess our own relationship to the natural world and renew, at the deepest level of personal integrity, a connection to it . . . The place to start is with faith, which for me is akin to a kind of spiritual gyroscope that spins in its own circumference in a stabilizing harmony with what is inside and what is out,' he preaches.[2]

He is in good company. Listen to some prominent advocates of spiritual ecology. 'Building an environmentally sustainable future depends on restructuring the global economy, major shifts in human reproductive behavior, and dramatic changes in values and lifestyles,' says Lester Brown, a leading American environmentalist. 'I very much doubt that we can heal the human spirit without discovering and learning to live by a new sense of purpose appropriate to the age and the ecological challenges we now face,' writes Jonathon Porritt, an equally prominent British environmentalist. 'Modern society will find no solution to the ecological problem unless it takes a serious look at its lifestyle . . . The seriousness of the ecological issue lays bare the depth of man's moral crisis,' opines the Pope. 'It is my own personal belief that we need to combine technological ability with, for want of a better description, spiritual readjustment and a realisation that certain truths are eternal,' urges the Prince of Wales.[3]

These are not modest aims. They are calls to change human nature. If such eco-optimism is well founded, then the argument of this book falls, and people are not calculating machines intricately designed to find cooperative strategies only when they assist enlightened self-interest. So if Chief Seattle really did live by his own philosophy of universal brotherhood with nature, I have a big explaining job to do. Ecologically noble savages – to borrow Rousseau's term – are inconsistent with the picture I have painted.

The chief's prescience, alas, is illusory. Nobody knows what he said that day. The only report, made thirty years later, was that he praised the generosity of the great white chief in buying his land. The entire 'speech' is a work of modern fiction. It was written for an ABC television drama by a screenwriter and professor of film, Ted Perry, in 1971. Though many environmentalists, Gore included, like to pretend otherwise, Chief Seattle was no tree-hugger. Among the few things we do know about him are that he was a slave owner and had killed almost all his enemies. As the case of Chief Seattle illustrates, the entire notion of living in harmony with nature is built on wishful thinking.[4]

Preaching and practice

Unless forcibly reminded of nature's cruelty, people tend to romanticize wildlife, seeing benevolence and overlooking viciousness. As George Williams has emphasized, crimes at least equivalent in their effects (if not their motives) to murder, rape, cannibalism, infanticide, deception, theft, torture and genocide are not just committed by animals, but are almost ways of life. Ground squirrels routinely eat baby ground squirrels; mallard drakes routinely drown ducks during gang rape; parasitic wasps routinely eat their victims alive from the inside; chimpanzees – our nearest relatives – routinely pursue gang warfare. Yet, as supposedly objective television programmes about nature repeatedly demonstrate, human beings just do not want to know these facts. They bowdlerize nature, desperately play up the slimmest of clues to animal virtue (dolphins saving drowning people, elephants mourning their dead), and clutch at straws suggesting that humankind somehow caused aberrant cruelty. When dolphins were recently found to be attacking porpoises off Scotland, animal 'experts' attributed this 'aberrant behaviour' to pollution of some kind, an assertion for which they admitted they had no evidence of any kind. We eliminate the negative and sentimentalize the positive.

We treat aboriginal people with the same condescending sentimentalism, as the enduring myth of the noble savage shows. But whereas in Rousseau's day the myth concerned social virtues, today it takes an ecological form. On an ethical level, respect for the sustainable use of the planet's resources has become one of the defining marks of a moral person. To express environmental sentiments is as politically correct today as to express any other bias in favour of the greater good: respect for minorities, distaste for crime and greed, belief in people's innate goodness, adherence to the golden rule. It is as disgraceful to be in favour of pollution these days as it was to be in favour of Satan in the thirteenth century. If, as I have argued in the preceding chapters, the human race is addicted to moralizing (though not necessarily acting) in favour of the greater good, for evolutionarily sound reasons, then it is no surprise that we seize upon political issues to express this instinct whenever we can. One of the most powerful ways to do so is to express the conservation ethic, to lament the fate of whales and rain forests, to disapprove of development, industry and growth – and to paint a rosy picture of how our ancestors (and our more tribal contemporaries) were morally better in this regard than we are.

This is, of course, hypocrisy. Just as we wish other people to turn the other cheek when hurt, but seek revenge on behalf of close relatives and friends, just as we urge morality far more than we act it, so environmentalism is something we prefer to preach than to practise. Everybody, it seems, wants a new road for themselves, but less road-building. Everybody wants another car, but wishes there were fewer on the road. Everybody wants two kids, but lower population growth.

The idea that native Americans had an environmental ethic that prevented their over-exploitation of nature is a recent invention of Westerners. When Daniel Day Lewis's screen father, Chingachgook, says to a deer his son has just killed, in the opening scene of the film *Last of the Mohicans*, 'We're sorry to kill you, brother. We do homage to your courage, speed and strength', he was being anachronistic. There is no evidence that the 'thank-you-dead-animal' ritual was a part of Indian folklore before the twentieth century. Even if it was a common practice, the animal was no less dead, however much the killer apologized.

The conventional wisdom holds that Indians were at one with nature, respecting and forbearing towards it, magically attuned to it and resolute in practising careful management so as not to damage the stock of their game. Archaeological sites throw doubt upon these comforting myths. Whereas wolves mostly kill old and very young animals, the elk killed by Indians were mostly in their prime. Cows were far more often killed than bulls, and very few elk lived to the same age that they do today. There is, concludes the ecologist Charles Kay, no evidence that native North Americans conserved big game.

Indeed, based on a comparison of the vegetation now and in the past, he argues that Indians were on the brink of driving the elk extinct in large parts of the Rockies before Columbus landed. While this extreme conclusion is disputed, certainly we know that throughout North America white men found game surprisingly scarce except on land disputed by two warring tribes (where the wars interfered with hunting). If there were spiritual and religious injunctions to conserve, they were remarkably ineffective. Indeed, Kay goes on to suggest, religious and shamanistic rituals may even have made things worse:

> Since native Americans saw no connection between their hunting and game numbers, the system of religious beliefs actually fostered the overexploitation of ungulate populations. Religious respect for animals does not equal conservation.[5]

Yet the myth persists, often for the explicit reason that preaching is seen as more important than practice. Even if it is not true of Amazon Indians, says one defender of Indian rights, that they conserve nature, it should continue to be said that they do, because 'any evidence of ecologically unsound activities by indigenous and traditional peoples undermines their basic rights to land, resources, and cultural practice'.[6]

The great Stone Age extinctions

The devastation wrought by our indigenous and traditional ancestors as they extinguished their way across the planet during and after the last ice age is only now becoming clear. Coincident with the first certain arrival of people in North America, 11,500 years ago, seventy-three per cent of the large mammal genera quickly died out. Gone were giant bison, wild horse, short-faced bear, mammoth, mastodon, sabre-toothed cat, giant ground sloth and wild camel. By 8,000 years ago, eighty per cent of the large mammal genera in South America were also extinct – giant sloths, giant armadillos, giant guanacos, giant capybaras, anteaters the size of horses.

This is known as the Pleistocene overkill. Sentimentalists among us still try to insist that it was a changing climate, not mankind, that did the damage, or that we only delivered the *coup de grâce* to species that were already in decline. It is remarkable how strong remains the wishful thinking for finding an excuse to believe in climatic change. But the sheer coincidence of the extinctions with the arrival of the first people, together with the fact that climate had often changed before as ice ages began and ended, and the strange selectivity of the extinguishing force – always killing the bigger animals – indict our species. There is also direct evidence: butchered carcasses with the spearheads of the Clovis people embedded in the bones. It is true that Africa and Eurasia saw no such sudden bursts of extinction of large mammals, and that mammoth hunting persisted for 20,000 years in Eurasia – but the mammoths and woolly rhinos went just as extinct there in the end as they did in North America. Besides, having lived with the human predator for millions of years already, the African and Eurasian fauna had already adjusted. The more vulnerable species had probably already died out, and the survivors had learnt to give us a wide berth, or to migrate in large herds. It is noticeable that the large North American mammals that did not go extinct in the Pleistocene overkill were mostly the ones that had crossed the land bridge from Asia together with people: the moose, elk, caribou, musk ox and brown bear. 'Did the animals simply fade away, or did we kill them?' asks Colin Tudge in *The Day Before Yesterday*; he answers his own question: 'Of course we killed them.'[7]

In other parts of the world, where people arrived suddenly and recently, the ecological effects of them were devastating – irrespective of climate. The guilt of the human species is not in doubt. Take Madagascar, where at least seventeen species of lemurs (all the diurnal ones larger than ten kilograms in weight, one as big as a gorilla), and the remarkable elephant birds – the biggest of which weighed 1,000 pounds – were dead within a few centuries of the island's first colonization by people in about 500 AD. It was a process repeated throughout the Pacific by the Polynesians and most spectacularly of all just six hundred years ago on New Zealand, where the first

Maoris sat down and ate their way through all twelve species of the giant moa birds (the biggest weighing a quarter of a ton) before turning cannibal in desperation. At one moa butchering site near Otago at least 30,000 were killed in a short time – and on average a third of the meat was left to rot, only the best haunches being taken. Entire ovens, with the roast haunches still in them, were left unopened, so abundant was the supply of meat. It was not just moas. Half of all New Zealand's indigenous land birds are extinct.

On Hawaii, we now know that there were about 100 species of unique Hawaiian birds, many of them large and flightless. Then, about 300 AD, a large mammal called humankind arrived. Within a short time no fewer than half of the Hawaiian birds were extinct. When this was first realized, after the excavation of an archaeological site in 1982, it was considered by native Hawaiians a major embarrassment for they had been arguing for many years that it was the arrival of Captain Cook that had upset a harmonious relationship between people and nature in the islands. In all, as the Polynesians colonized the Pacific, they extinguished twenty per cent of all the bird species on Earth.[8]

It took a little longer to wipe out Australia's large mammals. Yet soon after the arrival of the first people in Australia, possibly 60,000 years ago, a whole guild of large beasts vanished – marsupial rhinos, giant diprotodons, tree fellers, marsupial lions, five kinds of giant wombat, seven kinds of short-faced kangaroos, eight kinds of giant kangaroo, a two-hundred-kilogram flightless bird. Even the kangaroo species that survived shrank dramatically in size, a classic evolutionary response to heavy predation (which puts pressure on prey to start breeding when smaller).

It is crucial to remember that the fauna of the Americas, of Australia and of oceanic islands was naïve and unafraid of people. This, if anything, would have made conservation easier if the people had been so minded. Domestication or semi-domestication would have been simple. Consider this description of Lord Howe Island's virgin fauna when the first people reached it. In this case, unusually, the first people were sea-faring Europeans, the Polynesians having failed to find the island.

There was, wrote a member of the ship's party, '... A curious brown bird abt. the size of a Landrail in England walking totally fearless & unconcern'd in all part around us, so we had nothing more to do than to stand still a minute or two & knock as many as we pleased wt. a short stick – if you throwed at them and missed them, or even hit them with out killing them, they never made the least attempt to fly away ... The Pidgeons were also as tame as those already described & wd. sit upon the branches of trees till you might go and take them off with your hand ...'[9] Imagine a whole continent full of large mammals like that.

Yet our ancestors did not domesticate or manage the tame mammoths of North America or the trusting giant sloths of South America. They butchered them into oblivion. At Olsen-Chubbock, the site of ancient bison massacres in Colorado, where people regularly stampeded herds over a cliff, the animals lay in such heaps after a successful stampede that only the ones on top were butchered, and only the best joints were taken from them. Some conservationists![10]

Like a wolf on the fold

This ecological short-sightedness was not confined to hunters. In many parts of the world, ancient and simple-tooled people had astonishingly large effects on forests. In 1,000 years the Polynesians converted Easter Island, in the eastern Pacific, from a lush forest that provided wood for fishing canoes, food for many land birds and breeding sites for thirty kinds of seabirds, into a treeless, infertile and largely birdless grassland where famine, warfare and cannibalism thrived, and where vast stone statues lay abandoned in their quarries for want of logs to roll them into place. Petra, in Jordan, was once a thriving city in a thickly forested area, until the pressure of people turned it into a desert. The Mayan empire reduced the Yucatan peninsula to scrub and so fatally wounded itself. Chaco Canyon in New Mexico is the site of the largest building in North America before skyscrapers: it contained 650 rooms and 200,000 huge pine beams. Yet it was abandoned before the Spaniards arrived and its position is baffling. It is in a waterless desert with no pine trees for fifty miles or more. Archaeology has revealed that the Anasazi who built it had to go progressively farther for wood, eventually building a fifty-mile road specially to drag pine logs to the increasingly eroded and desiccated site. Finally they ran out of timber and their civilization fell. The forest has never recovered.[11]

History abounds with evidence that the limitations of technology or demand, rather than a culture of self-restraint, is what has kept tribal people from overexploiting their environment. Nor are the environmental practices of modern indigenous people as pretty a sight as romantic propaganda would have us believe. It is still routinely asserted that tribal people are careful to husband resources, careful to respect limits and careful to practise restraint, mediating these goals through religious and ritual observance. 'In my opinion,' writes Richard Nelson, 'the ethnographic record supports the existence of a widespread and well-developed tradition of conservation, land stewardship, and religiously based environmental ethics among Native Americans . . . we need to rediscover a deep, perhaps spiritually based, affiliation with life.'[12]

Practically every television programme about the tribal inhabitants of the rain forest repeats this assertion and its corollary that only recently and only in the West have people veered away from the tradition of living in spiritual harmony with nature. To take just one example, while writing this chapter I saw a programme about the Hoatzin bird in Ecuador, and heard the voice-over announce: 'Conserving a species for future use is a practical philosophy that all hunting peoples understand.'

Mysticism undoubtedly plays a large part in the life of tribal people. Some animals are thought to bring good luck; others bad luck. Complicated ceremonies may be performed before or after a hunt. Mountains are assumed to have feelings. Certain creatures are taboo, even if they appear to be edible. Sexual abstinence or fasting may precede an important hunt. True enough, all this, but does any of it work? As Hotspur remarks when the vainglorious Glendower claims that he can call spirits from the vasty deep: 'Why, so can I, or so can any man; But will they come when you do call for them?' Even if the religious ethic is towards conservation, people do not always live up to their ideals. Christianity preaches virtue, but few Christians are without sin. And even where the rituals do seem to favour conservation, coincidence, not intent, seems a better explanation.

For example, the Cree of Quebec rotate their hunting areas according to scapulimancy, the reading of runes on burnt caribou shoulder-blades. The shaman who reads the bones, remarkably, tells the hunters to avoid areas where game has been depleted by overhunting. Restraint is exercised and the game recovers. But a second's thought shows how flawed such an example is. Avoiding depleted areas makes sense anyway for the most selfish and straightforward of reasons – there is less to hunt. All the shaman does is pass on the information that he gathers from the hunters about which areas are depleted. The bones are irrelevant; they just add to the aura of professional indispensability, like the pompous language of a lawyer.

There have now been four studies of Amazon Indians that have directly tested their conservation ethic, by trying to find evidence that they practise systematic restraint in their hunting patterns to prevent the overexploitation of game. All four rejected the hypothesis. Ray Hames found that Yanomamo and Ye'kwana hunters spend more time in areas where there is more game. Since these areas are generally farther from the village, the hunters usually have to pass through depleted areas to reach these hunting grounds. If they were practising conservation, they would ignore any game they encountered on the way through the depleted area. But they do not. They always – without exception – pursue an animal they happen upon when in the depleted area, so long as it is big enough to be worth wasting effort and ammunition on.[13]

Michael Alvard found the same pattern in the Piro of Peru. With their shotguns (provided by the local priest) and bows and arrows, these Indians kill tapirs, peccaries, deer, capybara, spider monkeys, howler monkeys, agoutis and curassows. They too show a total lack of any systematic restraint in the depleted areas near the village, though they do ignore small game on the way out rather than waste their precious ammunition.[14]

William Vickers studied the Siona-Secoya of Ecuador for fifteen years, collecting records of 1,300 animal kills – the largest database on Amazon hunters ever collated. He recently reanalysed the data to look for evidence of a conservation ethic. He concluded that they did not practise conservation because they did not need to. Their population density was too low and their technology too limited to cause more than very local extinction. In that sense their practice was sustainable, but no thanks to their religious and ritual beliefs. A good shaman is supposed to remedy a shortage of game with spells, not tell the hunters to kill fewer animals. Only in recent years, under the pressure of white colonists and development, have they begun to think about the need to conserve the game in their shrinking forests. But they have done so rationally, not religiously. Conservation, says Vickers, is not a state of being but a rational response to new circumstances.[15]

Allyn MacLean Stearman found that the Yuqui of Bolivia are pure opportunists. They actually prefer to kill pregnant monkeys, or monkeys carrying young – because they are easier to catch and the foetus is considered a delicacy. They are casually cruel to wounded or captured animals. They fish with barbasco poison, which indiscriminately kills all the fish in a small pond or oxbow lake. And they are quite prepared to chop down whole trees to get ripe fruit (they used in days past to employ captured slaves to climb trees) with the result that fruiting trees are now scarce in some areas.[16]

The Rousseauian romantics prefer to believe that the Yuqui are aberrant in some way – bad Indians, rather than good Indians. But this is even more dangerous politics, says Stearman. It threatens to make Indian land rights contingent on their passing some test of ecological virtue, which is a test none should have to pass. 'We aren't nature lovers,' says Nicanor Gonzalez, a leader of the indigenous peoples movement. 'At no time have indigenous groups included the concepts of conservation and ecology in their traditional vocabulary.'[17]

The case of the Kayapo Indians is especially poignant. These inhabitants of central Brazil were interpreted by Rousseauian romantics as enlightened forest guardians. They were thought not only to protect, but also to create, patches of forest in the grasslands called apêtes as reserves for game and other valuable species. On the strength of this report they were granted a 20,000 square-mile reserve called Menkragnoti. The pop star Sting gave them $2 million towards its establishment. Within a few years they had begun an enthusiastic programme of selling concessions to gold miners and loggers.

The call to values

This is not to castigate Indians. It would be cheap and hypocritical of me indeed, sitting in my comfortable house dependent on immense quantities of fossil fuels and raw materials for my everyday needs, to be rude about an Indian just because he has found it necessary to sell some cheap logs for cash with which to buy necessities. He is endowed with vast reserves of knowledge about the natural history of his environment that I could never match – its dangers, its opportunities, its medicinal qualities, its seasons, its signs. He is a better conservationist than me in every conceivable way – simply by virtue of his material poverty. He leaves a smaller and more natural imprint on the planet. But this is because of the economic and technological limitations within which he lives, not because of some spiritual, inherent ecological virtue that he possesses. Give him the means to destroy the environment and he would wield them as unthinkingly as me – and probably with more efficiency.

So why do we destroy the environment? The answer is familiar. Environmental damage is caused by a form of the prisoner's dilemma, except that it is played by many players, not two. The problem in the prisoner's dilemma is to get two egoists to cooperate for the greater good, and to eschew the temptation to profit at the other's expense. Environmentalism is the same issue – how to prevent egoists producing pollution, waste and exhausted resources at the expense of more considerate citizens. For every time somebody exerts restraint, he only plays into the hands of a less considerate fellow human being. My forbearance is your opportunity, exactly as it was in the prisoner's dilemma, only this time the game is even harder to play because there are many players, not two.

Little wonder that environmentalists repeatedly and reflexively call for a change in human nature (or human values, as they prefer to call it). Fondly imagining that our instinctive egoism can be waved aside by persuasive calls to be good – as we saw in Chapter Seven, persuasive calls to be good are themselves a powerful human instinct; obeying them is not – they demand a new set of better values to live our lives by. To make this millennial cry more believable they point to how naturally ecological virtue seemed to come to our 'savage' ancestors. Like Rousseau they imagine that greed was invented just the other day, along with capitalism and technology. And they call for it to be disinvented as spiritual harmony with nature is reinvented.

Yet the conclusion that seems warranted is that there is no instinctive environmental ethic in our species – no innate tendency to develop and teach restrained practice. Environmental ethics are therefore to be taught in spite of human nature, not in concert with it. They do not come naturally. We all knew that anyway, did we not? Yet we persist in hoping that we'll find an ecological noble savage somewhere inside our breast to call out with the right chants and incantations. He's not in there. As Bobbi Low and Joel Heinen put it, 'Conservation philosophies relying on generalized and diffuse group benefits are probably doomed to failure, in the absence of individual or kinship benefits to conservation management. We would be delighted to be wrong, but suspect we are not.'[18]

But take courage! After all, the prisoner's dilemma turned out not to be the archetypal justification of human selfishness, but the reverse. Played repeatedly and discriminatingly, the game always favours the good citizen. Nice strategies like Tit-for-tat, Pavlov and Firm-but-fair win out over nasty ones. Perhaps game theory, too, can come to the rescue of the environmentalist's dilemma. Perhaps it can find a way for self-interested exploiters of the natural world to stop themselves killing the geese that lay golden eggs.

CHAPTER TWELVE

The Power of Property

In which governments are found wanting

The first man who, having enclosed a piece of land, thought of saying 'This is mine' and found people simple enough to believe him was the true founder of civil society. How many crimes, wars, murders; how much misery and horror the human race would have been spared if someone had pulled up the stakes and filled in the ditch and cried out to his fellow men: 'Beware of listening to this impostor. You are lost if you forget that the fruits of the earth belong to everyone and that the earth belongs to no one!'

Jean-Jacques Rousseau, *A Discourse on Inequality*, 1755

Give a man the secure possession of bleak rock, and he will turn it into a garden; give him nine years lease of a garden, and he will turn it into a desert . . . The magic of property turns sand into gold.

Arthur Young, *Travels*, 1787[1]

The rocky, fractal coast of Maine is ideal for lobsters. They swarm in the deep, cold-water inlets and off the coasts in considerable numbers. For hundreds of years they have been trapped and supplied as a delicacy to the rich of Boston and New York. In principle anybody can become a lobster fisherman. A licence is cheap and readily available from the state, so there are few legal barriers to entry. There is no limit on how many lobsters each fisherman can catch, so long as he does not kill breeding females or lobsters below a minimum size. The profits are good and the equipment is relatively simple.

All the ingredients are in place, in other words, for an environmental disaster. At the margin it will always pay a new fisherman to expand his effort even when the lobster stock cannot stand the pressure, for the old prisoner's dilemma reason – if he doesn't, somebody else will. And yet the lobstermen of Maine have at least until recently been thriving. They did not overfish the lobsters and have caught roughly the same quantity of lobsters – 16–22 million pounds a year – for fifty years. How did they avoid disaster?

The answer lies in a single phrase: property rights. In legal principle, as we have seen, anybody can trap lobsters anywhere. The fishing grounds are not privately owned. In practice, you would be well advised to think twice before setting up on your own. The whole coastline is divided into a series of territories, each of which 'belongs' to a particular 'harbour gang'. Although it is illegal to cut somebody's traps free from their buoys, it happens regularly to any intruder. Although there are no legal boundaries, each fisherman knows from landmarks on the shore exactly where he and other members of the gang must cease trapping. The territories are so precise that they can be mapped after a diligent questioning of the existing lobstermen.

The territories are jointly owned by the whole gang; there is no individual private property. If there were, the system would be unworkable, because lobsters move around at different seasons and a small territory that an individual could manage would be too small to be a reliable source of lobsters. Instead, the members of the gang move their traps at different seasons to different parts of the joint territory, which may cover 100 square miles.

Since the 1920s, however, there has been a gradual change in the way these gang territories are delineated, forced upon them by expanding populations and burgeoning technology, which makes it easier to stray across territory boundaries with impunity. Many territories are now defended only near the centre; towards the periphery they are a free-for-all. These 'nucleated territories' have smaller and fewer lobsters and the fishermen who use them make less money: $16,000 a year compared with $22,000 in the peripheral territories. The nucleated territories, in other words, are becoming an open-access fishery, and like all open-access fisheries, are beginning to show symptoms of overexploitation.

The extraordinary thing about the Maine lobster story, though, is not its deteriorating state, but that it has until now been so well run without any coercion or regulation by the state, and without individual private ownership (although with communal ownership).[2]

The rights of the commoner

Why so? The bleak message of the last chapter was that there is no such thing as ecological virtue; that the noble savage no more exists as an environmentalist than as Rousseau's fantasy. And yet lobster fishermen in Maine clearly do sustain the collective good. There seems to be a contradiction to be unravelled here.

A prisoner's dilemma played between many people is known as a 'tragedy of the commons'. When the Clovis people were on the way to exterminating the mammoths, imagine the folly of acting responsibly. If one individual said, 'No, I shall not kill that cow mammoth because she has a calf, and I must not damage the breeding stock', then how would he know the next Indian to come along would not think differently? How foolish he would look returning empty-handed to his hungry family, as another man brings meat from the very beast he rejected for his own family? Cooperation – i.e. restraint – by one party is opportunity for another. The rational individual would – did – kill the last two mammoths on the planet because he would know that another individual would get them if he did not.

This simple dilemma – the exact mirror-image of the problem of the provision of public goods, such as who will pay for the erection of a lighthouse (see Chapter Six) – has been known for ages, but the first person to put it in mathematical terms was Scott Gordon, an economist concerned with fisheries, in 1954. Gordon wrote thus:

Everybody's property is nobody's property. Wealth that is free for all is valued by none because he who is foolhardy enough to wait for its proper time of use will only find that it has been taken by another. The blade of grass that the manorial cowherd leaves behind is valueless to him, for tomorrow it may be eaten by another animal; the oil left under the earth is valueless to the driller, for another may legally take it; the fish in the sea are valueless to the fisherman, because there is no assurance that they will be there for him tomorrow if they are left behind today.[3]

The answer, said Gordon, was to privatize the resource or nationalize it and regulate its exploitation. In practice, only the latter course of action made sense for fisheries.

Fourteen years later, an authoritarian biologist named Garrett Hardin rediscovered this idea in preparing a lecture on population growth, and named it the tragedy of the commons, which term has stuck. Hardin's aim was not to try to solve the problem but to argue for the necessity of restrictions on the right to breed. 'Coercion,' he wrote, 'is a dirty word to most liberals now but it need not forever be so.'

To make his point, Hardin chose the example of medieval common land, which was widely believed to have been ruined by overgrazing, in comparison to enclosed land.

> The rational herdsman concludes that the only sensible course for him to pursue is to add another animal to his herd. And another; and another . . . But this is the conclusion reached by each and every rational herdsman sharing a commons. Therein lies the tragedy. Ruin is the destination toward which all men rush, each pursuing his own best interest in a society that believes in the freedom of the commons. Freedom in a commons brings ruin to all.[4]

In the abstract, this was true: free-for-alls are disastrously vulnerable to free-riding. The problem is, Hardin was wrong about grazing commons. Medieval commons were not disastrous free-for-alls. They were carefully regulated communal property, just like the lobster fisheries of Maine. True, there were relatively few written rights and not many obvious rules about who could graze them or cut coppice-wood on them. To an outsider, they looked like a free-for-all. But try adding your cattle to the common herd and you would soon discover the unwritten rules.

In practice, an English medieval common was a complex spider's web of jealously guarded property rights held under the supposedly benevolent umbrella of the lord of the manor, who owned the common but only on condition that he did not interfere with the rights of the commoners. There were rights of common of pasturage, estovers, turbary, pannage, piscary and common in the soil. Translated, these were rights to graze, cut wood, dig turf, turn out pigs to eat acorns, catch fish, or take gravel, sand or stone. And these rights were privately owned by individuals. As the manorial system broke down, commons came in effect to be owned jointly by those who possessed these rights in common, rights that were extinguished, converted or trampled upon in the process known as enclosure. But commons were never free-for-alls.[5]

To this day, many of the Pennine moors of the north of England retain the traditional medieval rule known as 'stinting'. Each sheep being grazed on the moor is free to go where it wishes, but the shepherd is not free to add any extra sheep. He possesses a certain number of 'stints', each of which entitles him to graze one ewe, and that sheep must be one that is born on the moor and 'hefted' to a flock already there (a hefted ewe is one that knows its place and stays within a short distance of the same spot all year; an unhefted ewe will wander). The number of stints is, in theory, calculated to ensure that the moor is not overgrazed. In the Middle Ages, most village commons were stinted this way. Now that stints are fully commoditized, subject to buying and selling for money, English commons are in effect partly privatized communal property. Much the same always applied to coppices in old English woodland: the rights to cut wood were owned. As Oliver Rackham, a historian of English forestry, has argued, 'commoners are no fools; they are well aware of the Hardin problem; they see the Tragedy coming and act to avert it; they draw up regulations to prevent overexploitation by any one shareholder. The court rolls of English commons make it clear that such regulations existed and could be updated to meet changing circumstances.'[6]

So it is nonsense to argue that just because something is communally owned it must suffer the tragedy of the commons. Common property and open-access free-for-alls are very different things. The old pre-enclosure English commons as a genuinely egalitarian place open to all is a nostalgic myth. Hardin was apparently unaware of this, and what he wrote was based on theory, not fact.[7]

Beware of nationalizers

Once this confusion is out of the way, it becomes obvious that all sorts of commons problems are readily and frequently managed in sensible, virtuous, sustainable ways by local people who entirely lack the pretensions to be trained economists. Conversely, it becomes obvious that it is the very trained experts who often undo, destroy and wreck sensible arrangements for managing commons. Elinor Ostrom, a political scientist, has been collecting examples of well-managed local commons for many years. In Japan and Switzerland, for example, she found forests that had been carefully looked after yet communally owned for many centuries.

On the Turkish coast, near the town of Alanya, there is a thriving inshore fishery. In the 1970s the local fishermen fell into the usual trap of overfishing, conflict and depletion. But they then developed an ingenious and complicated set of rules, allocating each known fishing location to each licensed fisher by lot in a pattern that rotates through the season. Enforcement is done by the fishers themselves, though the government recognizes the system in law. The fishery is now sustainable.

Near the Spanish city of Valencia, the waters of the River Turia are shared by more than 15,000 farmers in an arrangement that dates back at least 550 years and probably longer. Each farmer takes as much water as he needs from the distributory canal when his turn comes and wastes none. He is discouraged from cheating merely by the watchful eyes of his neighbours above and below him on the canal, and if they have a grievance they can take it to the *Tribunal de las Aguas*, which meets on Thursday mornings outside the Apostles' door of the Cathedral of Valencia. Records dating back to the 1400s show that cheating is rare. The *huerta* of Valencia is a profitable region, growing two crops a year. The region exported the system and its rules intact to New Mexico, where to this day self-governing irrigation systems thrive.[8]

In Almora, a hill district of Kumaon in northern India, made famous in the 1920s by the exploits of several man-eating tigers, there was a perfect example of how nationalizing a common creates, rather than solves, a tragedy of open access. In the 1850s the British government asserted absolute rights over all forest land in the district; it effectively nationalized land. The aim was to increase revenue for the government from the forests, ostensibly to the benefit of the local people. This was not peculiar to Almora; it was standard practice for the colonial government throughout India. The government banned trespass, felling, grazing and burning. The villagers resisted with growing militancy. For the first time they acted irresponsibly towards the forest, because it no longer belonged to them. A tragedy of the commons had been created.

By 1921, the problem was so severe that the government set up a Forest Grievance Committee, which re-communalized some of the forests under the Van Panchayat Act. Any two villagers or more could apply to the deputy commissioner of the district to create a Panchayat (or community forest) out of government-owned forest. The Panchayat council undertook to protect the forest from fires, encroachment, felling and cultivation, and to close twenty per cent of it to grazing every year. A study of six Panchayat forests in Almora conducted in 1990 concluded that three were well run and three poorly run. The three well-run ones were effective at monitoring their woodlands and fining rule breakers. They were considerably better at it than the central bureaucracy was in forests still belonging to the state.[9]

Another good example of the same phenomenon comes from northern Kenya. The Turkana people living along the Turkwell River near Lake Turkana once fed their goats upon the abundant acacia pods that fell from the riverside trees. From outside this looked like a free-for-all: all herdsmen used all the trees. But it was in fact not an open-access free-for-all, but a carefully regulated piece of private (communal) property. If anybody tried to let his animals browse a certain clump of trees without first negotiating permission from a committee of elders, he risked being driven off with sticks and, for a second offence, being killed. The government then intervened during a drought to regulate the browsing of the Turkwell trees. A new situation therefore developed in which a goat herder faced a genuine free-for-all; the government, not the elders, owned the trees. Tragically, and predictably, the trees were over-browsed and killed. Yet, bizarrely, so strong is the prejudice against private property among environmentalists, that the expert who described this case tried to make it an argument against privatization, not against nationalization.[10]

The tragedy of Leviathan

Hardin's legacy was to rehabilitate coercion by the state. It was a distinctly Hobbesian victory. Hobbes had argued in favour of a supreme sovereign power as the only way to enforce cooperation among its subjects. 'And covenants,' he wrote, 'without the sword, are but words, and of no strength to secure a man at all.' The only solution to tragedies of the commons, real or imagined, was seen in the 1970s as nationalization by the state. All across the world communal ownership, damned as inefficient by Hardin's logic, became an excuse for aggrandizement by governments. As one economist put it in 1973, shedding walrus tears, 'If we avoid the tragedy of the commons, it will *only* be by recourse to the tragic necessity of Leviathan.'[11]

This recipe was an unmitigated disaster. Leviathan creates tragedies of the commons where none were before. Consider the case of wildlife in Africa. All across the continent countries nationalized their game during colonial regimes and after independence in the 1960s and 1970s, arguing that it was the only way to prevent 'poachers' wiping out this commonly held resource. The result was that peasants now faced competition and damage from government-owned elephants and buffalo, and had no longer any incentive to look after the animals as a source of either meat or revenue. 'The African farmer's enmity towards elephants is as visceral as Western mawkishness is passionate,' said the head of the Kenya Wildlife Service, David Western. The decline of African elephants, rhinos and other animals is a tragedy of the commons, created by nationalization. This is proved by the fact that it has been spectacularly reversed wherever title to wildlife has been re-privatized to communities, such as in the Campfire programme of Zimbabwe in which sport hunters bid to buy the rights to kill game from committees of villagers. The villagers rapidly change their attitudes to the now-valuable game animals on their land. The acreage of private land devoted to wildlife has increased from 17,000 to 30,000 square kilometres since Zimbabwe granted title over wildlife to landowners.[12]

In irrigation systems in Asia, the damage done by government good intentions is even more striking. Irrigation systems in Nepal usually consist of a delicate bargain between the owners of the head-waters and the owners of fields farther downstream. By wasting water on thirsty crops like rice, or just being profligate, the upstream users can exhaust the available supply, leaving their downstream neighbours dry. Usually, however, they are more generous for purely self-interested reasons. Maintaining the diversion dams is hard work; the downstream users offer their labour in exchange for a fair share of the water. Consequently, when government steps in to build a permanent diversion dam, as it did at Kamala, the only effect was to upset an existing deal, remove the need for the upstream users to be good neighbours, and reduce the water that reached the downstream users. The project has been a spectacular failure. In contrast, where the government helped build some of the downstream branch canals, as it did at Pithuwa, there has been a coming together of users to create an efficient system of self-governing committees that allocate water, and the area served by the water has doubled.

In general, Nepalese irrigation systems run by the public sector average twenty per cent less crop yield than those run by the farmers, and are less equitable – less water reaching those at the downstream end. Concentrating the control of irrigation systems in bureaucratic hands has been a favourite game of governments since at least the pharaohs. It continued in colonial days, and is enthusiastically pursued to this day by aid agencies. It underestimates the ability of people to run their own systems, and overestimates the ability of bureaucrats. It creates a tragedy of the commons.[13]

Another case comes from the island of Bali, in Indonesia. Bali's landscape is man-made. Almost every accessible square inch has been terraced to make paddy fields. Sustainability, the ecological equivalent of virtue, is no problem. The farmers grow their own seed, and use no pesticides or fertilizer (blue-green algae fix nitrogen from the air). Rice has been grown in Bali since 1,000 BC and irrigation has been practised for almost as long. Irrigation tunnels and canals bring water from mountain lakes and streams down to the *subaks*, or farming villages, on the hillsides.

Irrigation is intimately connected with religion, each temple lying at a branch point in the canal network, and worship being apparently all about securing water by making offerings to upstream neighbours' temples. These temples dictate when each *subak* will have water to flood its fields, and when it must plant its rice. Traditionally, each *subak* plants all its fields at the same time and leaves all of its fields fallow at the same time.

Along, in the 1970s, came the Green Revolution in the form of the International Rice Research Institute, promulgating more vigorous strains of rice and promising the people better yields if they ceased to fallow their fields between crops. The result was disaster: water shortages and outbreaks of insect-borne viruses that ravaged the crops.

Why? Scientists were called in to find out. Stephen Lansing put the whole problem to his equivalent of a goddess (the computer) and it spake as follows. Before, everybody within each *subak* fallowed their crops at once, which destroyed the pest – it had nowhere to live during the fallow time. But each *subak* planted at different times, which ensured enough water for all. By interfering with simultaneous fallows and by creating sudden high demands for water from several areas, the Green Revolutionaries were spoiling a pattern that was far from being a mere hide-bound tradition. It was highly ingenious.

It was so ingenious that the person who worked it out must have been both clever and powerful. Who was he? The computer spake again. He was nobody. Order emerges perfectly from chaos not because of the way people are bossed about, but because of the way individuals react rationally to incentives. There is no omniscient priest in the top temple, just the simplest of conceivable habits. All it requires is that each farmer copies any neighbour who does better than he did. The result is synchrony within *subaks* and asynchrony between them. All without the slightest hint of central authority. Government, in the shape of rajahs or socialists, has done nothing to create the system; it only levies tax.[14]

Wherever you look, the reason for environmental troubles in the Third World turns out to be caused by the lack of clear property rights. Why do people mine the rain forest for logs when they could farm it for nuts and medicines? Because they can own the logs in a way that they cannot own them when they are trees. Why is Mexico exhausting its oil reserves more quickly, less efficiently and for less money than the United States? Because property rights to oil are better secured in America. The Peruvian economist Hernando de Soto argues that the poverty of the Third World is to be cured largely by creating secure property rights without which people have no chance to build their own prosperity. Government is not the solution to tragedies of the commons. It is the prime cause of them.[15]

Noble savage in the laboratory

There can be sustainability without Leviathan, after all. To prove this, Elinor Ostrom and his colleagues set up an experiment. They recruited eight students and gave each twenty-five tokens, which would be exchanged at the end of the study two hours later for real money. The students were given the chance of investing their tokens anonymously by computer in one of two markets. The first market gave a fixed rate of return, the same for every token invested. The second gave a different return according to the total number of tokens invested in it by all eight subjects. If only a small number of tokens were invested, the return was high, much higher than in the fixed-rate market. But the more tokens were invested in this second market, the lower the return until at a certain point you actually lost money by putting tokens into the second market.

This is designed to mimic a free-for-all environmental resource, like a fishery or a grazing meadow. A good return can be made if everybody exercises restraint, but the best return of all is made by he who does not exercise restraint when everybody else does – the free-rider. The question was, what would the students do? In the simplest version of the game, two hours of anonymous investing, as expected, the commons suffered from overgrazing. The students left with only twenty-one per cent of the maximum money they could have earned. Next the scientists gave the subjects a chance, half way through the session, to discuss the problem among themselves just once. They then went back to investing anonymously. The single discussion seemed to help. The return jumped to fifty-five per cent of the maximum available. Giving them the chance of repeated communication raised it still farther to seventy-three per cent. 'Mere jawboning', with no possibility of punishing free-riders, seemed to be remarkably effective at avoiding the tragedy.

By contrast, when Ostrom and her colleagues gave the subjects the chance to punish free-riders by fining them, but did not let them talk about it to agree a strategy, the return rate was low: thirty-seven per cent. Given the 'tax' cost to themselves of enforcing the fines, the true rate of return was only nine per cent. When they were allowed to communicate once and then develop their own method of fining the free-riders, the system began to work almost perfectly. The subjects walked away with ninety-three per cent of the maximum cash they could have earned. In such cases they came to agreements as to how many tokens each would be allowed to invest in the commons market and only four per cent of subjects defected from these agreements.[16]

So Ostrom's conclusions are that communication alone can make a remarkable difference to people's ability and willingness to exercise environmental restraint: indeed communication is more important than punishment. Covenants without swords work; swords without covenants do not. Take that, Hobbes! And so much for Hardin's plea for coercion.

If it moves, exploit it

This only makes the revelations of the last chapter more puzzling. In the absence of government interference, people are remarkably good at developing ways of solving the collective-action problem for environmental restraint among themselves, whether in a two-hour experiment in Indiana or a three-thousand-year experiment in Bali. So how is it that they so signally failed to stop themselves exterminating the megafauna of North and South America, Australia, New Guinea, Madagascar, New Zealand and Hawaii? How is it that the hunting practices of Amazon Indians bear not the slightest taint of effective ecological virtue?

The simplest answer is probably the right one. Animals move; irrigation systems do not. The key to solving common problems is the assertion of ownership – communal if necessary, individual if possible. Owning kangaroos or mastodons was as difficult as catching them. Even if a tribe were to deny hunting rights to outsiders within its territory, there is the dual problem of detecting trespassers and of preventing animals wandering into neighbouring territories. Or maybe there were perfectly adequate mechanisms for self-restraint among hunters in the Old World, which broke down in the excitement of discovering an abundant source of food in one of the new worlds. Did nobody among the early Maoris sit down after a moa feast and say, 'You know, if we go on like this, we're going to run out of moas to eat; maybe we should let a few alone to breed'? Evidently, if anybody did, nobody listened.

Evidence for the idea that people sustainably exploit only those things they can own comes from the fact that valuable living resources in tropical forests are generally treated with much more restraint if they do not move. Jared Diamond reports that New Guineans exhibit a conservation ethic only where individual rights are owned by individual people. A tree of a certain rare kind preferred for hollowing out as canoes belongs to he who finds it, and this rule is respected. The owner can therefore wait until he needs a new canoe before he fells it. Likewise, a tree used for display by certain birds of paradise is also privately owned by whoever finds it first. The owner has the sole right to shoot the birds for their prized, decorative plumes.[17]

The general rule that only nomadic or ephemeral resources are treated as free-for-alls and that the more static the resource the more privately it is owned is well illustrated by wildlife resources that are unusually static: exceptions that prove the rule. In North America, before the white men arrived, beavers were sustainably harvested by

Indians in many parts of the country. Near a beaver's dam there were marks upon the trees which revealed who owned the trapping rights to that particular beaver dam.

Or take the case of the megapodes, chicken-like birds of the islands of Australasia and the eastern Indies. Megapodes never incubate their eggs. Instead mostly they bury their eggs in specially constructed compost heaps to use the heat from rotting vegetation. Others dig burrows for their eggs in beaches to make use of the sun-warmed sand, or commute to volcanic islets to lay their eggs in burrows dug into sand warmed by geothermal activity below. One such geothermal beach in New Britain once attracted 53,000 birds. Not a single megapode ever sits on its eggs or looks after its chicks.

Large, protein-rich megapode eggs are a sought-after delicacy, and people compete for the right to harvest them. One person, or one community, usually owns the compost mound or the hot beach where the birds lay their eggs. This private ownership is crucial to the conservation of the birds. At one site on Haruku, a small island in the Moluccas, Rene Dekker recently discovered 5,000 pairs of megapodes laying eggs on one beach under the full moon. The harvesting rights were owned by one man, who paid an annual fee for the privilege and carefully left twenty per cent of the eggs to hatch. Other beaches are not now so lucky. The private-ownership system has broken down and a modern free-for-all has developed with disastrous results; eleven of the nineteen species of megapode are now under some kind of threat of extinction, mostly because of uncontrolled egg collecting.[18]

The difference between megapode nesting sites, beaver dams, bird-of-paradise trees and canoe trees on the one hand, and mammoths, tapir or herring on the other is that the former do not move. Property rights in the former are easily asserted, marked and defended. The thing that prevented our ancestors sustainably exploiting mammoths and elks was the fact that it was impossible to operate property rights in wild animals. These property rights need not be individual – they could be communal – but they were the key to ecological virtue.[19]

The hoarding taboo

The same conclusion applies to pollution and conservation in modern Western economies. Polluting companies adore regulation by government, because it protects them from civil suits and discourages new entrants to their business. They are terrified of environmental pressure from property rights asserted through the common law:

> Together trespass, nuisance and riparian rights have effectively empowered people to preserve or restore clean land, air and water – too effectively, apparently, for governments, which have worked assiduously to undermine property rights and the environmental protection they have fostered.[20]

Private property is often the friend of conservation; government regulation is often the enemy. Yet such a conclusion enrages environmentalists, who almost to a man and woman blame Western traditions of private property and greed for the damage that is being done to the environment, and recommend government intervention as the solution. There is, I believe, a simple reason for this. Private property or communal ownership by a small group is a logical response to a potential tragedy of the commons, but it is not an instinctive one. Instead, there is a human instinct, clearly expressed in hunter-gatherers, but present also in modern society, that strongly protests at any form of hoarding. Hoarding is taboo; sharing is mandatory. In Eskimos, anybody suspected of not sharing even his last cigarette is shamed into giving it up to the group. This hoarding taboo is the root of the common disapproval of private property. The Napoleonic code and the Hindu laws of partible inheritance, which enforce the division of property among many heirs, is part of this tradition. 'Property,' said the French anarchist Pierre-Joseph Proudhon, 'is theft.'

This is part of the almost obsessive egalitarianism of human beings, especially those in the hunter-gathering state. Anthropologists regularly report with surprise the way in which tribal people denigrate gifts, discussing how inadequate they are, or dismiss and ridicule the quality of a beast that has been killed by one of their fellows. As Elizabeth Cashdan has written of the !Kung people, 'If an individual does not minimize or speak lightly of his own accomplishments, his friends and relatives will not hesitate to do it for him . . . and if a person is not generous, the norms of sharing are "reinforced" by continued badgering and dunning for gifts.'[21]

In hunter-gatherer societies, nobody must be allowed to become big-headed; equality is all. We saw this in the way human coalitions, even more than chimpanzee ones, tame the ambitions of powerful individuals (see Chapter Eight). We see it again in the way hoarding is powerfully resented. But we also see that it is a constraint that is rapidly lifted as soon as some more sedentary and reliable way of life emerges that allows a powerful individual to rely on his own property rather than on the social insurance of sharing. Cashdan contrasts the egalitarian !Kung with the socially hierarchical //Gana, who rely for much of the year on predictable patches of wild melons that can be hoarded.

It is rare that the hoarding taboo survives in settled societies, but there are cases. Off the island of Manus, near New Guinea, lies a small sand cay, just two miles long and two hundred yards wide, but surrounded by a coral reef that stretches for eleven miles to the north. Ponam is its name, which is also the name of the tribe that inhabits it. In 1981 there were about 500 Ponams, of whom 300 still lived on the island. Apart from picking coconuts and keeping a few pigs, their main activity and source of food is to catch fish on the reef. This they do with spearguns and nets. The reef is parcelled up into communally owned fishing rights, each right belonging to a patrilineal clan. Canoes and nets are privately owned by the individual who made them. But to use his net, the owner must recruit a crew to help him. At the end of the day the catch is divided up into equal shares as follows: one share to each crew member, one share to the owner of the fishing right, one share to the owner of the canoe, one share to the owner of the net. However, nobody may take more than one share. If the three owners are one and the same person – if he who owns the boat also owns the net and the right – then he only gets one share. Those are the rules, and everybody obeys them. Only when the catch is huge does the owner take more than other people. When the catch is small, the owner usually forgoes his share altogether.

A more egalitarian system is hard to imagine. It rewards labour at the expense of capital, directing wealth away from those who have material possessions. This creates a strong disincentive to narrow the means of production: the larger the clan, the more labour it provides (rewarding) and the less capital (unrewarding). Like the Napoleonic code of inheritance, the Ponam customs reward communal ownership and discourage individual. It is an expression of the hoarding taboo.

But it is a wonder anybody ever makes a net or a canoe. When asked about this, the Ponams reply that they recognize the problem. When pressed they claim that the owner generally gets more fish – but they admit with further pressure that this is not true. They then claim that the owner gets an intangible reward: the esteem in which his clan is held goes up. The motive for ownership is social, not economic.[22]

Ponam is a parable for us all. Private ownership of wealth or property brings esteem and prestige, but it also brings envy and ostracism. Thus, however much we may recognize the arguments for property as a means to successful conservation of resources, we deeply dislike the argument. The modern conservationist finds himself in a cleft stick. Logic leads him to recommend private or communal property as the best means of giving people incentives to conserve. But his own hoarding taboo rebels against the idea. So he falls back on 'public ownership', comforting himself with the myth of perfect government. Observe the sleight of hand in this example:

> Most of Papua New Guinea (97%) is owned according to undocumented, customary tenure and only a very small proportion of PNG's spectacular landscapes, cultures and biological diversity is contained within legally-established protected areas. These unusual property rights, found only in countries of Oceania, restrict the ability of the Government to implement conservation measures through appropriation of land from traditional tenure to state control.[23]

If government were perfect, nationalization would work as well as such people hope. But government is imperfect, at least as much as markets are imperfect. It always diverts money to itself, whether corruptly or through Parkinson's Law. In addressing the environment, government is the cause of most problems, not the solution to them, precisely because it creates tragedies of the commons where none existed before. Would New Guineans cease to cut trees or shoot birds of paradise merely because they belonged to the government? Perhaps, if the government of New Guinea could afford fleets of helicopters hovering over the forest day and night with orders to shoot to kill. But that is hardly the government most of us want, or would even wish upon others.

Ecological virtue must be created from the bottom up, not the top down.[24]

CHAPTER THIRTEEN

Trust

In which the author suddenly and rashly draws political lessons

We do not suppose that the selfishness of human nature is ever to be overcome, but we would have the laws and institutions of society so framed as to give it all possible discountenance. *Morning Post*, January 1847

The *Post* probably imagines, because laws and institutions are intended to promote the public benefit, that they make society; but a different philosophy represents society as the natural product of the instincts of individuals.

Economist, January 1847[1]

Our minds have been built by selfish genes, but they have been built to be social, trustworthy and cooperative. That is the paradox this book has tried to explain. Human beings have social instincts. They come into the world equipped with predispositions to learn how to cooperate, to discriminate the trustworthy from the treacherous, to commit themselves to be trustworthy, to earn good reputations, to exchange goods and information, and to divide labour. In this we are on our own. No other species has been so far down this evolutionary path before us, for no species has built a truly integrated society except among the inbred relatives of a large family such as an ant colony. We owe our success as a species to our social instincts; they have enabled us to reap undreamt benefits from the division of labour for our masters – the genes. They are responsible for the rapid expansion of our brains in the past two million years and thence for our inventiveness. Our societies and our minds evolved together, each reinforcing trends in the other. Far from being a universal feature of animal life, as Kropotkin believed, this instinctive cooperativeness is the very hallmark of humanity and what sets us apart from other animals.

The evolutionary perspective is a long one. This book has in passing tried to nail some myths about when we adopted our cultured habits. I have argued that there was morality before the Church; trade before the state; exchange before money; social contracts before Hobbes; welfare before the rights of man; culture before Babylon; society before Greece; self-interest before Adam Smith; and greed before capitalism. These things have been expressions of human nature since deep in the hunter-gatherer Pleistocene. Some of them have roots in the missing links with other primates. Only our supreme self-importance has obscured this so far.

But self-congratulation is premature. We have as many darker as lighter instincts. The tendency of human societies to fragment into competing groups has left us with minds all too ready to adopt prejudices and pursue genocidal feuds. Also, though we may have within our heads the capacity to form a functioning society, we patently fail to use it properly. Our societies are torn by war, violence, theft, dissension and inequality. We struggle to understand why, variously apportioning blame to nature, nurture, government, greed or gods. The dawning self-awareness that this book has chronicled ought – indeed must – have some practical use. Knowing how evolution arrived at the human capacity for social trust, we can surely find out how to cure its lack. Which human institutions generate trust and which ones dissipate it?

Trust is as vital a form of social capital as money is a form of actual capital. Some economists have long recognized this. 'Virtually every commercial transaction has within itself an element of trust,' says the economist Kenneth Arrow. Lord Vinson, a successful British entrepreneur, cites as one of his ten commandments for success in business: 'Trust everyone unless you have a reason not to.' Trust, like money, can be lent ('I trust you because I trust the person who told me he trusts you'), and can be risked, hoarded or squandered. It pays dividends in the currency of more trust.

Trust and distrust feed upon each other. As Robert Putnam has argued, soccer clubs and merchant guilds have long reinforced trust in the successful north of Italy and fallen apart because of lack of trust in the more backward and hierarchical south. That is why two such similar peoples as the north Italians and the south Italians, equipped with much the same mixtures of genes, have diverged so radically simply because of a historical accident: the south had strong monarchies and godfathers; the north, strong merchant communities.[2]

Indeed, larger parallels spring to mind. Putnam argues that the North Americans developed a successful civic-minded society because they inherited a horizontally bonded version from the particular Britons who founded their cities, while the South Americans, stuck with the nepotism, authoritarianism and clientelism of medieval Spain, fell behind. You can take this too far. Francis Fukuyama argues unconvincingly that there is a broad difference between successful economies such as America and Japan and unsuccessful ones such as France and China because of the latter's addiction to hierarchical power structures. None the less, Putnam is indisputably on to something. Social contracts between equals, generalized reciprocity between individuals and between groups – these are at the heart of the most vital of all human achievements: the creation of society.[3]

The war of all against all

Much of this book has been a modern rediscovery – with added genetics and mathematics – of an age-old philosophical debate, a debate known by the name 'the perfectibility of man'. In various guises and at various times philosophers have argued that man is basically nice if he is not corrupted, or basically nasty if he is not tamed. Most famously, the debate pits Thomas Hobbes, on the side of nastiness, against Jean-Jacques Rousseau on the side of niceness.

Hobbes, though, was not the first to argue that man is a beast whose savage nature must be tamed by social contracts. Machiavelli said much the same two centuries before ('it must needs be taken for granted that all men are wicked,' he wrote). The Christian doctrine of original sin, refined by St Augustine, expressed a similar point: goodness comes as a gift from God. The Sophist philosophers of ancient Greece thought people inherently hedonistic and selfish. But it was Hobbes who made the argument political.[4]

Hobbes's intention, writing *Leviathan* in the 1650s in the wake of a century of religious and political civil war in Europe, was to argue that strong sovereign authority was required to prevent a state of perpetual fratricidal struggle. This was an unfashionable notion, for most seventeenth-century philosophers hewed to the ideal of a bucolic state of nature, typified in the supposedly peaceful and plentiful lives of American Indians, to justify their own search for a perfectly ordered society. Hobbes turned this on its head, arguing that the state of nature was one of war, not peace.[5]

Thomas Hobbes was Charles Darwin's direct intellectual ancestor. Hobbes (1651) begat David Hume (1739), who begat Adam Smith (1776), who begat Thomas Robert Malthus (1798), who begat Charles Darwin (1859). It was after reading Malthus that Darwin shifted from thinking about competition between groups to thinking about competition between individuals, a shift Smith had achieved in the century before.[6] The Hobbesian diagnosis – though not the prescription – still lies at the heart of both economics and modern evolutionary biology (Smith begat Friedman; Darwin begat Dawkins). At the root of both disciplines lies the notion that, if the balance of nature was not designed from above but emerged from below, then there is no reason to think it will prove to be a harmonious whole. John Maynard Keynes would later describe *The Origin of Species* as 'simply Ricardian economics couched in scientific language', and Stephen Jay Gould has said that natural selection 'was essentially Adam Smith's economics read into nature'. Karl Marx made much the same point: 'It is remarkable,' he wrote to Friedrich Engels in June 1862, 'how Darwin recognises among beasts and plants his own English society with its division of labour, competition, opening up of new markets, "inventions," and the Malthusian struggle for existence. It is Hobbes' *bellum omnium contra omnes*.'[7]

Darwin's disciple, Thomas Henry Huxley, chose exactly the same quotation from Hobbes to illustrate his argument that life is a pitiless struggle. For primitive man, he said, 'life was a continual free fight, and beyond the limited and temporary relations of the family, the Hobbesian war of each against all was the normal state of existence. The human species, like others, plashed and floundered amid the general stream of evolution, keeping its head above the water as it best might, and thinking neither of whence nor whither.' It was this essay that provoked Kropotkin to write *Mutual Aid*.

The argument between Huxley and Kropotkin had a personal edge. Huxley was a self-made man; Kropotkin an aristocratic revolutionary. Huxley was a meritocratic success with little time for dreamy outcast princes born in privilege; their falls from grace proved, to Huxley, their unfitness as surely as Huxley's own rise proved his fitness. 'It is open to us to try our fortune; and if we avoid impending fate, there will be a certain ground for believing that we are the right people to escape. *Securus judicat orbis*.'[8]

It was a short step from Huxley's meritocracy to the cruelty of eugenics. Evolution worked by sorting the strong from the weak, and it could be given a helping hand. Predestined not by their god but by their genes, the Edwardians came enthusiastically to the logical conclusion and began to sort the wheat from the chaff. Their successors in America and Germany committed the naturalistic fallacy, and sterilized and murdered millions of people in the belief they were thus improving the species or race. Although this project reached obscene depths under Hitler, it was widely supported, especially in the United States, by those on the left of the political spectrum, too. Indeed, Hitler was merely carrying out a genocidal policy against 'inferior', incurable or reactionary tribes that Karl Marx and Friedrich Engels had advocated in 1849 and that Lenin had begun to practise as early as 1918. It is even possible that Hitler got his eugenics not from Darwin or Spencer but from Marx, whom he read carefully when in Munich in 1913 and echoed closely on the topic. Many socialists were enthusiastic about eugenics, notably H. G. Wells who said, about 'black, and brown, and dirty-white, and yellow people who do not come into the new needs of efficiency' that they 'will have to go'.[9]

The Hobbesian search for a perfect society ended, therefore, in the gas chambers of Auschwitz, expressing not the human instinct for cooperation but the human instinct for genocidal tribalism, the Faustian bargain that comes, as we have seen, with groupishness.[10]

The noble savage

Hobbesian views prevailed in the century between 1845 and 1945. In the century before and the half century after, kinder and more Utopian views of human nature dominated political philosophy.

They, too, failed but not because they exploited the darker instincts of human beings. Instead, they mistakenly exaggerated the better instincts. And in a strange way, twice these Utopian ideals foundered in the South Pacific.

Of all the eighteenth-century Utopians, Jean-Jacques Rousseau was by far the most fanciful and by far the most influential. In his *Discourse on Inequality*, published in 1755, Rousseau painted a picture of humankind as a basically virtuous creature corrupted by civilization. Rousseau's idea of the noble savage, living in a harmonious state of nature until the invention of the evils of social life and property, was part daydream (Rousseau was awkward in grand society and resented it) and part polemic. For whereas Hobbes had wanted to justify authority after a period of anarchy, Rousseau wanted to undermine a corrupt, extravagant and potent monarchy that presided over, and taxed, a miserable populace. Until the invention of property and government, he argued, people had lived lives of freedom and equality. Modern society was a natural product of history, but it was decadent and ill. (Rousseau would have been at home in the modern environmental movement.)[11]

> Do not forget that society is natural to mankind as decrepitude is to man; that arts, laws, and governments are necessary to races as crutches are to the old; and the state of society being the extreme term at which men can arrive either sooner or later, it is not useless to show them the danger of going too quickly, and the miseries of a condition which they mistake for perfection.[12]

In 1768, when Rousseau's idea of the noble savage was at the height of its influence, Louis-Antoine de Bougainville discovered the island of Tahiti, named it New Cythera after the Peleponnesian isle where Aphrodite had first emerged from the sea, and compared it to the Garden of Eden. Despite Bougainville's own caution, his companions' description of the natives – beautiful, amorous, scantily clad, peaceable and wanting for nothing – caught the imagination of Paris, and of Rousseau's friend Denis Diderot in particular. Diderot wrote a fanciful supplement to Bougainville's account of his voyage in which a Tahitian sage expounds the virtues of their existence ('We are innocent; we are happy: and thou canst not but spoil our happiness. We follow the pure instinct of nature: thou hast sought to efface its character from our souls.') and a Christian chaplain is embarrassed by the generous sexual hospitality offered by Tahitian women.

James Cook visited Tahiti the next year and brought back similar reports of the plentiful, easy and strife-free life led by the islanders. They did not know shame, or hard work, or cold, or hunger. John Hawkesworth, commissioned to write up Cook's journal, laid it on thick, emphasizing especially the charms of the young ladies of Tahiti. Briefly, the South Seas were all the rage in art, in pantomime and poetry. The scorn of satirists like Samuel Johnson and Horace Walpole was ignored. The noble savage had been found in an eighteenth-century sexual fantasy.

Reaction was inevitable. On Cook's second voyage, the darker side of Tahitian life emerged: the human sacrifices, the regular practice of infanticide by a priestly caste, the vicious internecine quarrels, the rigid class hierarchy, the strict taboos on women eating in the presence of men, the incessant thievery practised by the natives upon the Europeans' possessions, the venereal disease – probably introduced by Bougainville's men. Jean François de Galaup, Comte de La Pérouse, who explored the Pacific and vanished in 1788, was especially hurt by his own disillusionment. Before he disappeared, he wrote bitterly: 'The most daring rascals of all Europe are less hypocritical than the natives of these islands. All their caresses were false.'[13] As the eighteenth century ended, with a French dictator waging war on the world and Parson Malthus persuading William Pitt that the poor laws only encouraged breeding and eventual famine, it was little wonder that the party was over in the South Pacific. Missionaries began to mobilize, intent on civilizing, or at least endowing with guilt, savages who now seemed more Hobbesian than Rousseauian.[14]

Paradise refound

History was to repeat itself in the South Seas. The twenty-three-year-old Margaret Mead went to Samoa in 1925 and returned, as Bougainville and Cook had returned from Tahiti nearly two hundred years before, with tales of a natural paradise free of the sins of the Western world, in which young men and young women lived easy, graceful, promiscuous lives largely free of the want, jealousy and violence that corrupted Western adolescence. Mead was a disciple of the anthropologist Franz Boas, who had reacted against an undue emphasis on eugenics in his native Germany. Boas, his face scarred from innumerable youthful duels, was not one to do things by halves. Instead of arguing that human behaviour was the product of both nature and nurture, he went to the other extreme, cultural determinism, and denied that anything but culture affected behaviour. To prove his point he needed to show the totipotency of human nature, the blank slate of John Locke. Given the right culture, he argued, we could create a society without jealousy, without love, without marriage, without hierarchy. Therefore humankind was infinitely malleable, and any Utopia was possible. To believe otherwise was irredeemably fatalist.

Mead was hailed for proving this to be more than wishful thinking. She brought back from Samoa apparently hard evidence of a society in which a different culture had produced a very different human nature. A culture of uninhibited free love among the Samoan youth prevented any adolescent angst, she argued. For fifty years Mead's Samoans stood as definitive proof of the perfectibility of man.[15]

But like Bougainville's Tahitian mirage, Mead's vanished on closer inspection. Whereas she had spent only five months in Manu'a, where her fieldwork was carried out, and only about twelve weeks of that on the research project Boas had required her to undertake, Derek Freeman spent over six years there in the 1940s and 1960s, and he discovered that Mead had been duped by her own wishful thinking and a mischievous streak in her informants. Observed without rose-tinted spectacles by Freeman, the Samoans could be like the Tahitians

Cook came to distrust on later visits, every bit as jealous, vicious and duplicitous as the rest of us. Virginity among unmarried adolescent girls was not a lightly held Christian novelty for free-loving Samoan women, but an ancient, respected cult whose violation had been punishable by death in pre-Christian days. Rape, far from being unknown, was so common that Samoa had one of the highest rates of rape in the world. Mead had let her Rousseauian preconceptions guide her, and she had missed the Hobbesian side.

Indeed, in 1987 one of Mead's chief informants came forward and admitted that she and her friend had as a prank hoaxed Mead with their accounts of their own supposedly flagrant promiscuity. As Freeman put it, 'Never can giggly fibs have had such far-reaching consequences' (although there was a precedent: the French traveller Labillardière was fooled in the eighteenth century by Tongans into reciting before the Academy of Sciences in Paris a string of phrases that he thought were Tongan numerals, but were actually obscenities).

The reaction of anthropologists to Freeman's revelation was itself the perfect refutation of Mead's creed. They reacted like a tribe whose cult had been attacked and shrine desecrated, vilifying Freeman in every conceivable way except by refuting him. If even cultural anthropologists, supposedly devoted to empirical truth and cultural relativism, act like a typical tribe, then there must be a universal human nature after all. They hold that there is no such thing as human nature independent of culture. They demonstrated that there is no such thing as culture independent of human nature. The slate is not blank after all.[16]

Margaret Mead committed, and many modern sociologists, anthropologists and psychologists continue to commit, a sort of reverse naturalistic fallacy. The naturalistic fallacy, identified by Hume and named by G. E. Moore, is to argue that what is natural is moral: deducing an 'ought' from an 'is'. Almost all biologists who speculate about the behaviour of bipedal apes are accused by the humanitarian establishment of committing this fallacy, even if they do not (many do). But the same establishment shows no embarrassment in continually and enthusiastically committing the reverse naturalistic fallacy: arguing from ought to is. Because something ought to be, then it must be. This logic is known today as political correctness, but it was shown in the drive launched by Boas, Benedict and Mead to argue that human nature must be infinitely malleable by culture because (they thought, wrongly) the alternative is fatalism, which is unacceptable.

Mead's creed spilled over into biology. Behaviourism held that animals' brains were black boxes which relied upon pure association to learn any task with equal ease. Its prophet, B. F. Skinner, wrote a science-fiction fantasy, *Walden Two*, about a world run by people like himself. 'We have no truck,' says Frazier, the founder of Walden Two, 'with philosophies of innate goodness – or evil either for that matter. But we do have faith in our power to change human nature.'[17]

Thus spake Lenin. The 1920s and 1930s, often seen as a time of lunatic obsession with genetic determinism, was also a time of lunatic obsession with environmental determinism: the belief that man could be remade entirely into new man just by education, propaganda and force. Under Stalin this Lockean faith in changing nature was even applied to wheat. Trofim Lysenko argued, and those who gainsaid him were shot, that wheat could be made more frost-hardy not by selection but by experience. Millions died hungry to prove him wrong. The inheritance of acquired characteristics remained an official doctrine of Soviet biology until 1964. Unlike the genetic determinism of Hitler, Stalin's environmental variety went on to infect other peoples.[18]

In her extraordinary autobiographical account of the Chinese revolution, *Wild Swans*, Jung Chang gives the perfect example of why Communism failed because it failed to change human nature. In 1949 her mother married a young Communist official, who repeatedly refused to use his position to help her or other members of his family. He would not let her share his car on a long journey which she undertook on foot, lest it seem like favouritism; he refused to pardon a convicted counter-revolutionary guerrilla who had saved her life, because, he argued, the man had tipped her off precisely in the hope of currying favour with her husband; he demoted her by two grades in the party hierarchy just to forestall any suggestion that she had been given a higher rank than was justified; he intervened to veto his own elder brother's promotion in a tea-marketing business; again and again he refused to show a normal preference for his family, because he put the revolution first and believed that to be nice to your relatives was to discriminate against your non-relatives. He was right. Communism would have worked if there were more such men, though it would have been a bleak kind of success in which people could not be nice to their relatives. But most people are not like Wang Shou-yu. Indeed, given their immunity from criticism, Communist officials have consistently proved more corruptible and more nepotistic than democratic ones. Universal benevolence evaporates on the stove of human nature.[19]

As Herbert Simon has put it, 'In our century we have watched two great nations, the People's Republic of China and the Soviet Union, strive to create a "new man," only to end up by acknowledging that the "old man" – perhaps we should say the "old person" – self-interested and concerned with his or her economic welfare, or the welfare of the family, clan, ethnic group, or province, was still alive and well.'[20]

Fortunately, there proved to be, in Lionel Trilling's words, 'a residue of human quality beyond cultural control.' Otherwise, Russians would now be irredeemably corrupted people, which they plainly are not. Karl Marx designed a social system that would only have worked if we were angels; it failed because we were beasts. Human nature had not been changed at all. 'I would rather hope [that man has some innate nature] than be stuck with a human tabula rasa on which any tyrants or do-gooders can write their (always benign) messages at will. And I think man has such a nature, that it is intensely social, and that it gives the lie to all sanctimonious manipulators from Mill through Stalin,' said Robin Fox.[21]

Who stole the community?

If the refashioning of society by competitive struggle led to the gas chambers, and the refashioning of society by cultural dogma led to the horrors of Mao's Cultural Revolution, then would it not be safer to abandon all ideas of importing science into politics altogether? Perhaps so. Certainly, I am not going to fall into the trap of pretending that our dim and misty understanding of the human social instinct can be instantly translated into a political philosophy. For a start, it teaches us that Utopia is impossible, because society is an uneasy compromise between individuals with conflicting ambitions, rather than something designed directly by natural selection itself.

None the less, the new 'gene-tilitarian' understanding of human instincts that this book has explored leads to a few simple precepts for avoiding mistakes. Human beings have some instincts that foster the greater good and others that foster self-interested and anti-social behaviour. We must design a society that encourages the former and discourages the latter.

Consider, for example, a glaring paradox of free enterprise. If we declare that Smith, Malthus, Ricardo, Friedrich Hayek and Milton Friedman are right, and that man is basically motivated by self-interest, do we not by that very declaration encourage people to be more selfish? By recognizing the inevitability of greed and self-interest, we seem to approve it.

The essayist William Hazlitt certainly believed so, fulminating in his 'Reply to Malthus' that:

> It is neither generous nor just to come in the aid of the narrow prejudices and hard-heartedness of mankind, with metaphysical distinctions and the cobwebs of philosophy. The balance inclines too much on that side already, without the addition of false weights.[22]

In other words, the reason we must not say that people are nasty is that it is true. More than 150 years later, Robert Frank discovered that economics students, after being taught that people were essentially self-interested, grew more so themselves: they defected in prisoner's dilemma games more than other students. The real Ivan Boesky and the fictitious Gordon Gecko (in the film *Wall Street*) both notoriously eulogized greed. 'Greed is all right, by the way,' said Boesky in his commencement address at the University of California at Berkeley, in May 1986. 'I want you to know that. I think greed is healthy. You can be greedy and still feel good about yourself.' Spontaneous applause broke out.[23]

It has become almost axiomatic that injunctions such as this are responsible for the breakdown of community spirit of recent years. Taught in the 1980s, against our better natures, to be selfish and greedy we have dropped our civic responsibilities and caused our societies to descend into amorality. This is the standard, soft-left explanation for rising crime and insecurity.

So the first thing we should do to create a good society is to conceal the truth about humankind's propensity for self-interest, the better to delude our fellows into thinking that they are noble savages inside. It is a distasteful idea for those of us who think the truth is more interesting than lies, however white. But the distaste need not worry us for long, because the white-lying is already happening. As we have repeatedly encountered in this book, propagandists always exaggerate the niceness of people, partly to flatter them and partly because the message is more palatable. People wish to believe in noble savages. As Robert Wright has argued:

> The new [selfish gene] paradigm strips self-absorption of its noble raiment. Selfishness, remember, seldom presents itself to us in naked form. Belonging as we do to a species (*the* species) whose members justify their actions morally, we are designed to think of ourselves as good and our behaviour as defensible, even when these propositions are objectively dubious.[24]

Only those politicians who enjoy saying unpopular things will rock this particular boat. Said Margaret Thatcher, notoriously and scandalously: 'There's no such thing as society. There are individual men and women, and there are families.'

Of course, Thatcher had a serious point. At the core of her philosophy was the idea that if you fail to recognize the basic opportunism of human beings, then you fail to notice how government is composed of self-interested individuals rather than saints who only work for the greater good. Government is then just a tool for interest groups and budget-maximizing bureaucrats to bid up each other's power and reward at the expense of the rest of us. It is not a neutral, motiveless machine for delivering social benefits. She was against government's inherent corruption, rather than its ideals.

And yet Thatcher and her allies were articulating what is, in some ways, the most Rousseauian argument – that government does not impose virtue on inherently evil people, but corrupts the original virtue of the market place. Her mentor, Friedrich Hayek, appealed to a golden age when the noble savage was free of all regulation: without regulation from the state there would not be chaos but prosperity.[25]

Time magazine, profiling Newt Gingrich as its man of the year in December 1995, made the point succinctly:

> Here's the way the world used to work: Liberals believed human beings, if not perfectible, were at least subject to improvement . . . Conservatives believed human beings were fundamentally flawed . . . Here's the way the world works today: conservatives believe . . . human beings aren't evil; the government is. Liberals, on the other hand, believe conservatives are dangerous romantics . . . They are ready to believe some souls are inherently evil and beyond redemption.[26]

If my argument in this book is right, then the conservatives are not such dangerous romantics, because the human mind contains numerous instincts for building social cooperation and seeking a reputation for niceness. We are not so nasty that we need to be tamed by intrusive government, nor so nice that too much government does not bring out the worst in us, both as its employees and as its clients.

So let us examine the individualists' case: that government is the problem, not the solution. The collapse of community spirit in the last few decades, and the erosion of civic virtue, is caused in this analysis not by the spread and encouragement of greed but by the dead hand of Leviathan. The state makes no bargain with the citizen to take joint responsibility for civic order, engenders in him no obligation, duty or pride, and imposes obedience instead. Little wonder that, treated like a naughty child, he behaves like one.

As Putnam's Italian example shows, where authority replaces reciprocity, the sense of community fades. In Britain, the welfare state and the mixed-economy 'corpocracy' replaced thousands of effective community institutions – friendly societies, mutuals, hospital trusts and more, all based on reciprocity and gradually nurtured virtuous circles of trust – with giant, centralized Leviathans like the National Health Service, nationalized industries and government quangos, all based on condescension. Because more money was made available through higher taxes, something was gained at first. But soon the destruction wrought to Britain's sense of community was palpable. Because of its mandatory nature the welfare state encouraged in its donors a reluctance and resentment, and in its clients not gratitude but apathy, anger or an entrepreneurial drive to exploit the system. Heavy government makes people more selfish, not less.[27]

I hold to no foggy nostalgia that the past was any better. Most of the past was a time of authority, too – the hierarchical authority of a feudal, aristocratic or industrial system. (It was also, of course, a time of less material prosperity, but that is down to inferior technology, not inferior government.) The medieval vassal and the factory worker had no freedom to build trust and reciprocity between equals either. I am not contrasting the present with the past. But I do believe that there have been glimpses of a better way, of a society built upon voluntary exchange of goods, information, fortune and power between free individuals in small enough communities for trust to be built. I believe such a society could be more equitable, as well as more prosperous, than one built upon bureaucratic statism.

I live close to one of the great old cities of Britain, Newcastle upon Tyne. In two centuries it has been transformed from a hive of enterprise and local pride, based on locally generated and controlled capital and local mutual institutions of community, into the satrapy of an all-powerful state, its industries controlled from London or abroad (thanks to the collectivization of people's savings through tax relief for pension funds), and its government an impersonal series of agencies staffed by rotating officials from elsewhere whose main job is to secure grants from London. Such local democracy as remains is itself based entirely on power, not trust. In two centuries the great traditions of trust, mutuality and reciprocity on which such cities were based have been all but destroyed – by governments of both stripes. They took centuries to build. The Literary and Philosophical

Society of Newcastle, in whose magnificent library I researched some of this book, is but a reminder of the days when the great inventors and thinkers of the region, almost all of them self-made men, were its ambitious luminaries. The city is now notorious for shattered, impersonal neighbourhoods where violence and robbery are so commonplace that enterprise is impossible. Materially, everybody in the city is better off than a century ago, but that is the result of new technology, not government. Socially, the deterioration is marked. Hobbes lives, and I blame too much government, not too little.

If we are to recover social harmony and virtue, if we are to build back into society the virtues that made it work for us, it is vital that we reduce the power and scope of the state. That does not mean a vicious war of all against all. It means devolution: devolution of power over people's lives to parishes, computer networks, clubs, teams, self-help groups, small businesses – everything small and local. It means a massive disassembling of the public bureaucracy. Let national and international governments wither into their minimal function of national defence and redistribution of wealth (directly – without an intervening and greedy bureaucracy). Let Kropotkin's vision of a world of free individuals return. Let everybody rise and fall by their reputation. I am not so naïve as to think this can happen overnight, or that some form of government is not necessary. But I do question the necessity of a government that dictates the minutest details of life and squats like a giant flea upon the back of the nation.

For St Augustine the source of social order lay in the teachings of Christ. For Hobbes it lay in the sovereign. For Rousseau it lay in solitude. For Lenin it lay in the party. They were all wrong. The roots of social order are in our heads, where we possess the instinctive capacities for creating not a perfectly harmonious and virtuous society, but a better one than we have at present. We must build our institutions in such a way that they draw out those instincts. Pre-eminently this means the encouragement of exchange between equals. Just as trade between countries is the best recipe for friendship between them, so exchange between enfranchised and empowered individuals is the best recipe for cooperation. We must encourage social and material exchange between equals for that is the raw material of trust, and trust is the foundation of virtue.

Sources and Notes

PROLOGUE

1 Woodcock, George and Avakumovic, Ivan. 1950. *The Anarchist Prince: A Biographical Study of Peter Kropotkin.* T. V. Boardman and Co. London; Kropotkin, Peter. 1902/1972. *Mutual Aid: A Factor in Evolution.* Allen Lane, London.
2 Kropotkin. *Mutual Aid. op. cit.*

CHAPTER ONE

1 Hölldobler, B. and Wilson, E. O. 1990. *The Ants.* Harvard University Press, Cambridge, Mass.
2 Gould, S. J. 1978. *Ever Since Darwin.* Burnett Books, New York.
3 Gordon, D. M. 1995. The development of organization in an ant colony. *American Scientist* 83:50–57.
4 Buss, L. W. 1987. *The Evolution of Individuality.* Princeton University Press, Princeton.
5 Bonner, J. T. 1993. *Life Cycles: Reflections of an Evolutionary Biologist.* Princeton University Press, Princeton; Dawkins, R. 1996. *Climbing Mount Improbable.* Viking, London.
6 Sherman, P. W., Jarvis, J. U. M. and Alexander, R. D. 1991. *The Biology of the Naked Mole Rat.* Princeton University Press, Princeton. Perhaps the most remarkable thing about naked mole rats is that Richard Alexander predicted their existence. Knowing nothing about them he postulated, by analogy with termites, a burrowing social mammal in 1976. The social life of the naked mole rat became clear soon after.
7 The idea that life is gradually coagulating into larger and larger teams does not imply that smaller life forms will disappear. But it does mean that more and more of the small life forms will adopt parasitic habits as more and more of the sun's energy flows through the big forms of life.

8 Dawkins, R. 1982. *The Extended Phenotype*. Freeman, Oxford.
9 Kessin, R. H. and Van Lookeren Campagne, M. M. 1992. The development of a social amoeba. *American Scientist* 80:556–65.
10 Maynard Smith, J. and Szathmary, E. 1995. *The Major Transitions in Evolution*. W. H. Freeman, Oxford.
11 Paradis, J. and Williams, G. C. 1989. *Evolution and Ethics: T. H. Huxley's Evolution and Ethics with New Essays on its Victorian and Sociobiological Context*. Princeton University Press, Princeton.
12 Hamilton, W. D. 1964. The genetical evolution of social behaviour. I, II. *Journal of Theoretical Biology* 7:1–52.
13 Hamilton, W. D. 1996. *Narrow Roads of Gene Land. Vol. 1: Evolution and Social Behaviour*. W. H. Freeman/Spektrum, Oxford.
14 Dawkins, R. 1976. *The Selfish Gene*. Oxford University Press, Oxford
15 Hamilton. *Narrow Roads of Gene Land. Vol. 1. op. cit.*
16 Hamilton. The genetical evolution of social behaviour. *op. cit.*; Williams, G. C. 1966. *Adaptation and Natural Selection: A Critique of Some Current Evolutionary Thought*. Princeton University Press, Princeton; Williams, G. C. 1992. *Natural Selection*. Oxford University Press, Oxford; Dawkins. *The Selfish Gene. op. cit.*

Curiously, it was a poem called 'The Fable of the Bees', published in 1714 by an English cynic and satirist, that first glimpsed this possibility. Bernard Mandeville's poem was a defence of the necessity of vice. Just as hunger is necessary if we are to eat and thrive, he argued, so selfish ambition is necessary if we are to prosper and reap public goods. The practice of pure benevolence is incompatible with the development of a prosperous commercial society. Mandeville, B. 1714/1755. *The Fable of the Bees: or Private Vices, Public Benefits*. 9th edn. Edinburgh.

17 Sen, A. K. 1977. Rational fools: a critique of the behavioral foundations of economic theory. *Philosophy and Public Affairs* 6:317–44. See also Hirshleifer, J. 1985. The expanding domain of economics. *American Economic Review* 75:53–68.
18 Note that Haig's idea of conflict in pregnancy does not imply any conscious decision to fight on the part of either mother or offspring. It only implies an evolved physiological mechanism designed by selection to achieve these effects.
19 Haig, D. 1993. Genetic conflicts in human pregnancy. *Quarterly Review of Biology* 68:495–531; D. Haig, interviews.
20 Ratnieks, F. L. W. 1988. Reproductive harmony via mutual policing by workers in eusocial hymenoptera. *American Naturalist* 132:217–36; Oldroyd, B. P., Smolenski, A. J., Cornuet, J.-M. and Crozier, R. H. 1994. Anarchy in the beehive. *Nature* 371:749.

21 Matsuda, H. and Harada, Y. 1990. Evolutionarily stable stalk to spore ratio in cellular slime molds and the law of equalization of net incomes. *Journal of Theoretical Biology* 147:329–44.
22 Buchanan, J. M. 1969. *Cost and Choice*. Markham Publishing, Chicago; Buchanan, J. M. and Tullock, G. 1982. *Towards a Theory of the Rent-Seeking Society*. A. & M. Press, Texas.
23 Parkinson's Law first appeared in an anonymous article in the *Economist*, 19 November 1955, pages 635–7. It was later expanded by Parkinson into a book. See also Nozick, R. 1974. *Anarchy, State and Utopia*. Basic Books, New York.
24 Robinson, W. S. 1913. *A Short History of Rome*. Rivingtons, London. Shakespeare gives Menenius a similar speech in *Coriolanus*.
25 Nesse, R. M. and Williams, G. C. 1995. *Evolution and Healing: The New Science of Darwinian Medicine*. Weidenfeld and Nicolson, London. The book was entitled *Why We Get Sick* in its American edition.
26 Charlton, B. G. 1995. Endogenous parasitism: a biological process with implications for senescence. *Evolutionary Theory* (in press).
27 Leigh, E. G. 1991. Genes, bees and ecosystems: the evolution of a common interest among individuals. *Trends in Evolution and Ecology* 6:257–62.
28 Buss, L. W. 1987. *The Evolution of Individuality*. Princeton University Press, Princeton.
29 I am indebted to David Haig for the information that human beings have B chromosomes at the rate of 2–3 per cent of live births.
30 Bell, G. and Burt, A. 1990. B-chromosomes: germ-line parasites which induce changes in host recombination. *Parasitology* 100:S19–S26. The parasitic nature of B chromosomes was suspected as long ago as 1945: Stergren, G. 1945. Parasitic nature of extra fragment chromosomes. *Botaniska Notiser* (1945):157–63.
31 Leigh, E. G. 1971. *Adaptation and Diversity*. Freeman, Cooper, San Francisco.

CHAPTER TWO

1 Wilson, D. S. and Sober, E. 1994. Reintroducing group selection to the human and behavioral sciences. *Behavioral and Brain Sciences* 17:585–654. Note also that the Hutterite fission process is a perfect illustration of John Rawls's thought experiment in developing his theory of justice. A just society, argued Rawls, would be one that you would draw up when a veil of ignorance concealed the specific role you would play in that society. See

Rawls, J. 1972. *A Theory of Justice*. Oxford University Press, Oxford; and Dennett, D. 1995. *Darwin's Dangerous Idea*. Simon and Schuster, New York.

2 Paradis, J. and Williams, G. C. 1989. *Evolution and Ethics: T. H. Huxley's Evolution and Ethics with New Essays on its Victorian and Sociobiological Context*. Princeton University Press, Princeton.

3 Alexander, R. D. 1987. *The Biology of Moral Systems*. Aldine de Gruyter, Hawthorne, New York.

4 Layton, R. H. 1989. Are sociobiology and social anthropology compatible? The significance of sociocultural resources in human evolution. In: *Comparative Socioecology* (eds. Standen, V. and Foley, R.) Blackwell, Oxford.

5 Selfish means doing things for me; altruistic means doing things for you; groupish means doing things for us. Margaret Gilbert made this useful distinction in her commentary on Wilson and Sober. Reintroducing group selection. *op. cit.*

6 Franks, N. R. and Norris, P. J. 1987. Constraints on the division of labour in ants: D'Arcy Thompson's Cartesian transformations applied to worker polymorphism. *Experientia Supplementum* 54:253–70.

7 Szathmary, E. and Maynard Smith, J. 1995. The major evolutionary transitions. *Nature* 374:227–32.

8 West, E. G. 1990. *Adam Smith and Modern Economics*. Edward Elgar Publishing, Vermont.

9 Maynard Smith, J. and Szathmary, E. 1995. *The Major Transitions in Evolution*. W. H. Freeman, Oxford.

10 Bonner, J. T. 1993. Dividing labour in cells and societies. *Current Science* 64:459–66.

11 Stigler, G. J. 1951. The division of labor is limited by the extent of the market. *Journal of Political Economy* 59:185–93.

12 Ghiselin, M. T. 1978. The economy of the body. *American Economic Review* 68 (2):233–7.

13 Ghiselin, M. T. 1974. *The Economy of Nature and the Evolution of Sex*. University of California Press, Berkeley.

14 Smith, A. 1776/1986. *The Wealth of Nations*. Penguin, Harmondsworth.

15 Brittan, S. 1995. *Capitalism with a Human Face*. Edward Elgar, Aldershot.

16 Buss, L. W. 1987. *The Evolution of Individuality*. Princeton University Press, Princeton.

17 Coase, R. H. 1976. Adam Smith's view of man. *Journal of Law and Economics* 19:529–46.

18 Emerson, A. C. 1960. The evolution of adaptation in population systems.

In: *Evolution after Darwin. Vol. 1.* (ed. Tax, S.). University of Chicago Press, Chicago.
19 K. Hill and H. Kaplan, personal communication.
20 Spindler, K. 1993. *The Man in the Ice.* Weidenfeld and Nicolson, London.
21 Smith. *The Wealth of Nations. op. cit.*; Wright, R. 1994. *The Moral Animal.* Pantheon, New York.

CHAPTER THREE

1 Rousseau, J.-J. 1755/1984. *A Discourse on Inequality.* Penguin, Harmondsworth.
2 Hofstadter, D. 1985. *Metamagical Themas: Questing for the Essence of Mind and Pattern.* Basic Books, New York. See also Dennett, D. 1995. *Darwin's Dangerous Idea.* Simon and Schuster, New York.
3 P. Hammerstein, personal communication.
4 Poundstone, W. 1992. *Prisoner's Dilemma: John von Neumann, Game Theory and the Puzzle of the Bomb.* Oxford University Press, Oxford.
5 Rapoport, A. and Chummah, A. M. 1965. *Prisoner's Dilemma.* University of Michigan Press, Ann Arbor.
6 Maynard Smith, J. and Price, G. R. 1973. The logic of animal conflict. *Nature* 246:15–18. In the original paper, the term 'dove' was changed at the last minute to 'mouse' in deference to George Price's religious sensibilities.
7 Rapoport, A. 1989. *The Origins of Violence.* Paragon House, New York.
8 Axelrod, R. 1984. *The Evolution of Cooperation.* Basic Books, New York.
9 Trivers, R. L. 1971. The evolution of reciprocal altruism. *Quarterly Review of Biology* 46:35–57.
10 The best book on game theory in biology is: Sigmund, K. 1993. *Games of Life.* Oxford University Press, Oxford.
11 Wilkinson, G. S. 1984. Reciprocal food sharing in the vampire bat. *Nature* 308:181–4. Recent research confirms that even the more transient and less family-obsessed male vampire bats reciprocate in the same fashion: see DeNault, L. K. and McFarlane, D. A. 1995. Reciprocal altruism between male vampire bats, *Desmodus rotundus. Animal Behaviour* 49:855–6.
12 Cheney, D. L. and Seyfarth, R. M. 1990. *How Monkeys See the World.* Chicago University Press, Chicago.
13 Trivers. The evolution of reciprocal altruism. *op. cit.*

CHAPTER FOUR

1 R. Barton, personal communication.
2 Dunbar, R. 1996. *Grooming, Gossip and the Evolution of Language*. Faber and Faber, London.
3 Heinsohn, R. and Packer, C. 1995. Complex cooperative strategies in group-territorial African lions. *Science* 269:1260–62.
4 Martinez-Coll, J. C. and Hirshleifer, J. 1991. The limits of reciprocity. *Rationality and Society* 3:35–64.
5 Binmore, K. 1994. *Game Theory and the Social Contract. Vol. 1: Playing Fair*. MIT Press, Cambridge, Mass.
6 Badcock, C. 1990. Three fundamental fallacies of modern social thought. *Sociological Notes No. 5*. The referee's remarks were quoted by Lyall Watson in the *Financial Times*, 15 July 1995.
7 Recently, new prisoner's dilemma games have been played in space, rather than time and, if anything, they reinforce the impression that Tit-for-tat is a powerful strategy. See Hutson, V. C. L. and Vickers, G. T. 1995. The spatial struggle of tit-for-tat and defect. *Philosophical Transactions of the Royal Society of London* B 348:393–404; Ferriere, R. and Michod, R. E. 1995. Invading wave of cooperation in a spatially iterated prisoner's dilemma. *Proceedings of the Royal Society of London* B 259:77–83.
8 Nowak, M. A., May, R. M. and Sigmund, K. 1995. The arithmetics of mutual help. *Scientific American* 272:50–55.
9 Boyd, R. 1992. The evolution of reciprocity when conditions vary. In: *Coalitions and Alliances in Humans and Other Animals* (eds. Harcourt, A. H. and de Waal, F. B. M.). Oxford University Press, Oxford.
10 Kitcher, P. 1993. The evolution of human altruism. *Journal of Philosophy* 90:497–516.
11 Frank, R. H., Gilovich, T. and Regan, D. T. 1993. The evolution of one-shot cooperation. *Ethology and Sociobiology* 14:247–56.

CHAPTER FIVE

1 Barrett, P. H., Gantrey, P. J., Herbert, S., Kohn, D. and Smith, S. (eds.) 1987. *Charles Darwin's Notebooks, 1836–1844*. Cambridge University Press, Cambridge.
2 Friedl, E. 1995. Sex the invisible. *American Anthropologist* 96:833–44. The Ik of Uganda are a partial exception to this rule: they are secretive

about meals because of near starvation. See Turnbull, C. 1972. *The Mountain People*. Simon and Schuster, New York.
3 Fiddes, N. 1991. *Meat: A Natural Symbol*. Routledge, New York.
4 Galdikas, B. 1995. *Reflections of Eden: My Life with the Orang-utans of Borneo*. Victor Gollancz, London.
5 Stanford, C. B., Wallis, J., Mpongo, E. and Goodall, J. 1994. Hunting decisions in wild chimpanzees. *Behaviour* 131:1–18; Tutin, C. E. G. 1979. Mating patterns and reproductive strategies in a community of wild chimpanzees (*Pan troglodytes schweinfurthii*). *Behavioral Ecology and Sociobiology* 6:29–38.
6 Hawkes, K. 1995. Foraging differences between men and women. In: *The Archaeology of Human Ancestry* (eds. Steele, J. and Shennan, S.). Routledge, London.
7 Ridley, M. 1993. *The Red Queen: Sex and the Evolution of Human Nature*. Viking, London.
8 Kimbrell, A. 1995. *The Masculine Mystique*. Ballantine Books, New York.
9 *Economist*, 5 March 1994, p. 96.
10 Berndt, C. H. 1970. Digging sticks and spears, or the two-sex model. In: *Woman's role in Aboriginal society. Australian Aboriginal Studies, No. 36* (ed. Gale, F.). Australian Institute of Aboriginal Studies, Canberra; Megarry, T. 1995. *Society in Prehistory*. Macmillan, London.
11 Steele, J. and Shennan, S. (eds.) 1995. *The Archaeology of Human Ancestry*. Routledge, London.
12 Bennett, M. K. 1954. *The World's Food*. Harper and Row, New York. Cited in Fiddes. *Meat: A Natural Symbol. op. cit.*
13 De Waal, F. B. M. 1989. Food sharing and reciprocal obligations among chimpanzees. *Journal of Human Evolution* 18:433–59.
14 Hill, K. and Kaplan, H. 1989. Population and dry-season subsistence strategies of the recently contacted Yora of Peru. *National Geographic Research* 5:317–34.
15 Winterhalder, B. 1986. Diet choice, risk and food-sharing in a stochastic environment. *Journal of Anthropological Archaeology* 5:369–92.

CHAPTER SIX

1 The fantasy of the hegemony of grass was developed by Calder, N. 1984. *Timescale: An Atlas of the Fourth Dimension*. Chatto and Windus, London.
2 Leakey, R. E. 1994. *The Origin of Humankind*. Weidenfeld and Nicolson, London.
3 Guthrie, R. D. 1990. *Frozen Fauna of the Mammoth Steppe: The Story*

of Blue Babe. University of Chicago Press, Chicago; Zimov, S. A., Churprynin, V. I., Oreshko, A. P., Chapin, F. S., Reynolds, J. F. and Chapin, M. C. 1995. Steppe–tundra transition: a herbivore-driven biome shift at the end of the Pleistocene. *American Naturalist* 146:765–94.
4 Farmer, M. F. 1994. The origin of weapon systems. *Current Anthropology* 35:679–81; C. Keckler, interview.
5 Hawkes, K. 1993. Why hunter-gatherers work: an ancient version of the problem of public goods. *Current Anthropology* 34:341–61.
6 Blurton-Jones, N. G. 1987. Tolerated theft, suggestions about the ecology and evolution of sharing, hoarding and scrounging. *Social Science Information* 26:31–54.
7 Hill, K. and Kaplan, H. 1994. On why male foragers hunt and share food. *Current Anthropology* 34:701–6.
8 Winterhalder, B. 1996. A marginal model of tolerated theft. *Ethology and Sociobiology* 17:37–53.
9 Alexander, R. D. 1987. *The Biology of Moral Systems*. Aldine de Gruyter, Hawthorne, New York.
10 Brealey, R. A. and Myers, S. C. 1991. *Principles of Corporate Finance*. 4th edn. McGraw Hill, New York.
11 Wilson, J. Q. 1993. *The Moral Sense*. The Free Press, New York.
12 Sahlins, M. 1966/1972. *Stone Age Economics*. Aldine de Gruyter, Hawthorne, New York.
13 Alasdair Palmer. Do you sincerely want to be rich? *Spectator*, 5 November 1994, p. 9.
14 Zahavi, A. 1995. Altruism as a handicap – the limitations of kin selection and reciprocity. *Journal of Avian Biology* 26:1–3.
15 Cronk, L. 1989. Strings attached. *The Sciences*, May–June 1989:2–4.
16 Davis, J. 1992. *Exchange*. Open University Press, Buckingham.
17 Benedict, R. 1935. *Patterns of Culture*. Routledge and Kegan Paul, London.
18 *ibid*.
19 Davis. *Exchange*. *op. cit.*

CHAPTER SEVEN

1 Nesse, R. 1994. Commentary in Wilson, D. S. and Sober, E. 1994. Reintroducing group selection to the human and behavioral sciences. *Behavioral and Brain Sciences* 17:585–654.
2 Cosmides and Tooby worried that including the word altruistic might

confuse those who did not know what it meant; but they tried 'selfless' instead and got much the same result.

3 Barkow, J., Cosmides, L. and Tooby, J. 1992. *The Adapted Mind*. Oxford University Press, Oxford.

4 L. Sugiyama, talk to the Human Behavior and Evolution Society meeting, Santa Barbara, June 1995.

5 L. Cosmides, interview.

6 Stephen Budiansky suggested this point to me.

7 Barkow, Cosmides and Tooby. *The Adapted Mind. op. cit.*

8 Trivers, R. L. 1971. The evolution of reciprocal altruism. *Quarterly Review of Biology* 46:35–57.

9 Ghiselin, M. T. 1974. *The Economy of Nature and the Evolution of Sex*. University of California Press, Berkeley. The point about Christianity has been well made by the newspaper columnist Matthew Parris.

10 Frank, R. H. 1988. *Passions within Reason*. Norton, New York.

11 The blue-tit story comes from Birkhead, T. R. and Moller, A. P. 1992. *Sperm Competition in Birds: Evolutionary Causes and Consequences*. Academic Press, London.

12 Trivers. The evolution of reciprocal altruism. *op. cit.*; Trivers, R. L. 1983. The evolution of a sense of fairness. In: *Absolute Values and the Creation of the New World. Vol. 2*. The International Cultural Foundation Press, New York.

13 Frank. *Passions within Reason. op. cit.*

14 Binmore, K. 1994. *Game Theory and the Social Contract. Vol. 1: Playing Fair*. MIT Press, Cambridge, Mass.

15 Alexander, R. D. 1987. *The Biology of Moral Systems*. Aldine de Gruyter, Hawthorne, New York; Singer, P. 1981. *The Expanding Circle: Ethics and Sociobiology*. Farrar, Straus and Giroux, New York.

16 V. Smith, posting on the HBES list, E-mail, June 1995. Talk to the Human Behavior and Evolution Society meeting, Santa Barbara, June 1995.

17 Frank. *Passions within Reason. op. cit.*

18 Kagan, J. 1984. *The Nature of the Child*. Basic Books, New York.

19 D. Cheney, talk at the Royal Society, 4 April 1995.

20 Wilson, J. Q. 1993. *The Moral Sense*. Free Press, New York.

21 Damasio, A. 1995. *Descartes's Error: Emotion, Reason and the Human Brain*. Picador, London.

22 Dawkins, R. 1976. *The Selfish Gene*. Oxford University Press, Oxford.

23 Jacob Viner, quoted in Coase, R. H. 1976. Adam Smith's view of man. *Journal of Law and Economics* 19:529–46.

CHAPTER EIGHT

1 Packer, C. 1977. Reciprocal altruism in olive baboons. *Nature* 265: 441–3.

2 Noe, R. 1992. Alliance formation among male baboons: shopping for profitable partners. In: *Coalitions and Alliances in Humans and Other Animals* (eds. Harcourt, A. H. and de Waal, F. B. M.). Oxford University Press, Oxford.

3 Van Hooff, J. A. R. A. M. and van Schaik, C. P. 1992. Cooperation in competition: the ecology of primate bonds. In: *Coalitions and Alliances* (eds. Harcourt and de Waal). *ibid.*

4 Silk, J. B. 1992. The patterning of intervention among male bonnet macaques: reciprocity, revenge and loyalty. *Current Anthropology* 33:318–25; Silk, J. B. 1993. Does participation in coalitions influence dominance relationships among male bonnet macaques? *Behaviour* 126:171–89; Silk, J. B. 1995. Social relationships of male bonnet macaques. *Behaviour* (in press).

5 Dennett, D. 1995. *Darwin's Dangerous Idea.* Simon and Schuster, New York.

6 Pinker, S. 1994. *The Language Instinct.* Allen Lane, London.

7 Cronin, H. 1991. *The Ant and the Peacock.* Cambridge University Press, Cambridge; Rawls, J. 1972. *A Theory of Justice.* Oxford University Press, Oxford.

8 Nishida, T., Hasegawa, T., Hayaki, H., Takahata, Y. and Uehara, S. 1992. Meat-sharing as a coalition strategy by an alpha male chimpanzee? In: *Topics in Primatology. Vol. 1: Human Origins* (eds. Nishida, T., McGrew, W. C., Marler, P., Pickford, M. and de Waal, F. B. M.). Tokyo University Press, Tokyo.

9 De Waal, F. B. M. 1982. *Chimpanzee Politics.* Johns Hopkins University, Baltimore; de Waal, F. B. M. 1992. Coalitions as part of reciprocal relations in the Arnhem chimpanzee colony. In: *Coalitions and Alliances* (eds. Harcourt and de Waal). *op. cit.*; de Waal, F. B. M. 1996. *Good Natured: The Origins of Right and Wrong in Humans and Other Animals.* Harvard University Press, Cambridge, Mass.

10 Boehm, C. 1992. Segmentary 'warfare' and the management of conflict: comparison of East African chimpanzees and patrilineal-patrilocal humans. In: *Coalitions and Alliances* (eds. Harcourt and de Waal). *op. cit.*

11 Connor, R. C., Smolker, R. A. and Richards, A. F. 1992. Dolphin alliances and coalitions. In: *Coalitions and Alliances* (eds. Harcourt and de Waal). *op. cit.*

12 Boehm, Segmentary 'warfare' and the management of conflict. In: *Coalitions and Alliances* (eds. Harcourt and de Waal). *op. cit.*
13 Moore, J. In Wilson, D. S. and Sober, E. 1994. Reintroducing group selection to the human and behavioral sciences. *Behavioral and Brain Sciences* 17:585–654; Alexander, R. D. 1987. *The Biology of Moral Systems.* Aldine de Gruyter, Hawthorne, New York; Trivers, R. L. 1983. The evolution of a sense of fairness. In: *Absolute Values and the Creation of the New World. Vol.* 2. International Cultural Foundation Press, New York.
14 Boehm. Segmentary 'warfare' and the management of conflict. In: *Coalitions and Alliances* (eds. Harcourt and de Waal). *op. cit.*
15 Gibbon, E. 1776–88/1993. *The History of the Decline and Fall of the Roman Empire. Vol. IV.* Everyman, London.

CHAPTER NINE

1 Mesterson-Gibbons, M. and Dugatkin, L. A. 1992. Cooperation among unrelated individuals: evolutionary factors. *Quarterly Review of Biology* 67:267–81; Rissing, S. and Pollock, G. 1987. Queen aggression, pheometric advantage and brood raiding in the ant *Veromessor pergandei. Animal Behaviour* 35:975–82; Hölldobler, B. and Wilson, E. O. 1990. *The Ants.* Harvard University Press, Cambridge, Mass.
2 Wynne-Edwards, V. C. 1962. *Animal Dispersion in Relation to Social Behaviour.* Oliver and Boyd, London.
3 Lack, D. 1966. *Population Studies of Birds.* Clarendon Press, Oxford.
4 Hamilton, W. D. 1971. Geometry for the selfish herd. *Journal of Theoretical Biology* 31:295–311; Alexander, R. D. 1989. Evolution of the human psyche. In: *The Human Revolution* (eds. Mellars, P. and Stringer, C.). Edinburgh University Press, Edinburgh.
5 Szathmary, E. and Maynard Smith, J. 1995. The major evolutionary transitions. *Nature* 374:227–32; Alexander, R. D. 1987. *The Biology of Moral Systems.* Aldine de Gruyter, Hawthorne, New York.
6 Boyd, R. and Richerson, P. 1990. Culture and cooperation. In: *Beyond Self-Interest* (ed. Mansbridge, J. J.). Chicago University Press, Chicago.
7 R. Boyd, talk to the Royal Society, 4 April 1995.
8 Boyd and Richerson. Culture and cooperation. In: *Beyond Self-Interest* (ed. Mansbridge). *op. cit.*
9 Sutherland, S. 1992. *Irrationality: The Enemy Within.* Constable, London.
10 Ridley, M. 1993. *The Red Queen: Sex and the Evolution of Human Nature.* Viking, London. See also Hirshleifer, D. 1995. The blind leading the blind: social influence, fads and informational cascades. In: *The New*

Economics of Behaviour (ed. Tommasi, M.). Cambridge University Press, Cambridge; Bikhchandani, S., Hirshleifer, D. and Welch, I. 1992. A theory of fads, fashion, custom and cultural change as informational cascades. *Journal of Political Economy* 100:992–1026.
11 Hirshleifer. The blind leading the blind. In: *The New Economics of Behaviour* (ed. Tommasi). *op. cit.*; Bikhchandani, Hirshleifer, and Welch. A theory of fads. *Journal of Political Economy op. cit.*
12 Simon, H. 1990. A mechanism for social selection of successful altruism. *Science* 250:1665–8.
13 Soltis, J., Boyd, R. and Richerson, P. J. 1995. Can group-functional behaviors evolve by cultural group selection? An empirical test. *Current Anthropology* 36:473–94.
14 C. Palmer, talk to the Human Behavior and Evolution Society, Santa Barbara, June 1995.
15 John Hartung, correspondence.
16 Lyle Steadman, personal communication.
17 W. McNeill, address to the Human Behavior and Evolution Society, Ann Arbor, Michigan, August 1994.
18 Richman, B. 1987. Rhythm and Melody in Gelada vocal exchanges. *Primates* 28:199–223; Storr, A. 1993. *Music and the Mind*. HarperCollins, London.
19 Gibbon, E. 1776–88/1993. *The History of the Decline and Fall of the Roman Empire. Vol. I.* Everyman, London.
20 Mead is quoted in Bloom, H. 1995. *The Lucifer Principle*. Atlantic Monthly Press, Boston; Alexander. *The Biology of Moral Systems. op. cit.*.
21 Hartung, J. 1995. Love thy neighbour. *The Skeptic*, Vol. 3, No. 4; Keith, A. 1947. *Evolution and Ethics*. G. P. Putnam's Sons, New York.

CHAPTER TEN

1 Sharp, L. 1952. Steel axes for Stone-Age Australians. *Human Organisation*, Summer 1952:17–22.
2 I am grateful to Kim Hill for making this point to me.
3 Layton, R. H. 1989. Are sociobiology and social anthropology compatible? The significance of sociocultural resources in human evolution. In: *Comparative Socioecology* (eds. Standen, V. and Foley, R.). Blackwell, Oxford.
4 Chagnon, N. 1983. *Yanomamo, the Fierce People*. 3rd edn. Holt, Rinehart and Winston, New York.
5 Benson, B. 1989. The spontaneous evolution of commercial law. *Southern*

Economic Journal 55:644–61; Benson, B. 1990. *The Enterprise of Law.* Pacific Research Institute, San Francisco.
6 Coeur was jailed on the island of Chios and died there in 1456. His magnificent Gothic palace is one of the principal sights of Bourges.
7 Watson, A. M. 1967. Back to gold – and silver. *Economic History Review*, 2nd Series, 20:1–34.
8 Samuelson, P. Quoted in Brockway, G. P. 1993. *The End of Economic Man.* Norton, New York, p. 299.
9 Heilbronner, R. L. 1961. *The Worldly Philosophers.* Simon and Schuster, New York.
10 Sraffa, P. (ed.) 1951. *The Works of David Ricardo.* Cambridge University Press, Cambridge.
11 Roberts, R. D. 1994. *The Choice: A Fable of Free Trade and Protectionism.* Prentice Hall, Englewood Cliffs, New Jersey.
12 Alden-Smith, E. 1988. Risk and uncertainty in the 'original affluent society': evolutionary ecology of resource sharing and land tenure. In: *Hunters and Gatherers. Vol 1: History, Evolution and Social Change* (eds. Ingold, T., Riches, D. and Woodburn, J.). Berg, Oxford.
13 Robert Layton, interview; Paul Mellars, talk to Royal Society; Gamble, C. 1993. *Timewalkers: The Prehistory of Global Colonisation.* Alan Sutton, London.

CHAPTER ELEVEN

1 Gore, A. 1992. *Earth in the Balance: Ecology and the Human Spirit.* Houghton Mifflin, Boston.
2 *ibid.*
3 Brown, L. 1992. *State of the World.* Worldwatch Institute, Washington, DC; Porritt, J. 1991. *Save the Earth.* Channel Four Books, London; the Pope is quoted in: Gore. *Earth in the Balance. op. cit.*; the Prince of Wales wrote the foreword to Porritt's book.
4 Kauffman, W. 1995. *No Turning Back: Dismantling the Fantasies of Environmental Thinking.* Basic Books, New York; Budiansky, S. 1995. *Nature's Keepers: The New Science of Nature Management.* Weidenfeld and Nicolson, London.
5 Kay, C. E. 1994. Aboriginal overkill: the role of the native Americans in structuring western ecosystems. *Human Nature* 5:359–98.
6 Posey, D. W. 1993. Quoted in Vickers, W. T. 1994. From opportunism to nascent conservation. The case of the Siona Secoya. *Human Nature* 5:307–37.

7 Tudge, C. 1996. *The Day Before Yesterday*. Jonathan Cape, London; Stringer, C. and McKie, R. 1996. *African Exodus*. Jonathan Cape, London.
8 Steadman, D. W. 1995. Prehistoric extinctions of Pacific island birds: biodiversity meets zooarcheology. *Science* 267:1123–31.
9 Flannery, T. 1994. *The Future Eaters*. Reed, Chatswood, New South Wales.
10 Alvard, M. S. 1994. Conservation by native peoples: prey choice in a depleted habitat. *Human Nature* 5:127–54.
11 Diamond, J. 1991. *The Rise and Fall of the Third Chimpanzee*. Radius Books, London.
12 Nelson, R. 1993. Searching for the lost arrow: physical and spiritual ecology in the hunter's world. In: *The Biophilia Hypothesis* (eds. Kellert, S. R. and Wilson, E. O.). Island Press, Washington, DC.
13 Hames, R. 1987. Game conservation or efficient hunting? In: *The Question of the Commons* (eds. McCay, B. and Acheson, J.). University of Arizona Press, Tucson.
14 Alvard. Conservation by native peoples. *Human Nature*. *op. cit.*
15 Vickers, W. T. 1994. From opportunism to nascent conservation. The case of the Siona-Secoya. *Human Nature* 5:307–37.
16 Stearman, A. M. 1994. 'Only slaves climb trees': revisiting the myth of the ecologically noble savage in Amazonia. *Human Nature* 5:339–57.
17 Quoted in *ibid.*
18 Low, B. S. and Heinen, J. T. 1993. Population, resources and environment. *Population and Environment* 15:7–41.

CHAPTER TWELVE

1 Quoted in Brubaker, E. 1995. *Property Rights in the Defence of Nature*. Earthscan, London.
2 Acheson, J. 1987. The lobster fiefs revisited. In: *The Question of the Commons* (eds. McCay, B. and Acheson, J.). University of Arizona Press, Tucson.
3 Gordon, H. S. 1954. The economic theory of a common-property resource: the fishery. *Journal of Political Economy* 62:124–42.
4 Hardin, G. 1968. The tragedy of the commons. *Science* 162:1243–8.
5 'Commons'. Booklet produced by the Country Landowners' Association, October 1992. No. 16/92.
6 Townsend, R. and Wilson, J. A. 1987. In: *The Question of the Commons* (eds. McCay and Acheson). *op. cit.*; Oliver Rackham, correspondence with the author.

7 To be fair to Hardin, he has since said that he should, more accurately, have used the term 'unmanaged commons' in his original article.
8 Ostrom, E. 1990. *Governing the Commons: The Evolution of Institutions for Collective Action*. Cambridge University Press, Cambridge; Brown, D. W. 1994. *When Strangers Cooperate: Using Social Conventions to Govern Ourselves*. The Free Press, New York.
9 Ostrom, E., Gardner, R. and Walker, J. 1993. *Rules, Games and Common-pool Resources*. Princeton University Press, Princeton.
10 Monbiot, G. 1994. The tragedy of enclosure. *Scientific American* January 1994:140.
11 Ophuls, W. 1973. Leviathan or oblivion. In: *Towards a Steady-state Economy* (ed. Daly, H. E.) Freeman, San Francisco.
12 Bonner, R. 1993. *At the Hand of Man*. Knopf, New York; Sugg, I. and Kreuter, U. P. 1994. *Elephants and Ivory: Lessons from the Trade Ban*. Institute of Economic Affairs, London.
13 Ostrom, E. and Gardner, R. 1993. Coping with asymmetries in the commons: self-governing irrigation systems can work. *Journal of Economic Perspectives* 7:93–112.
14 S. Lansing, talk to the Human Behavior and Evolution Society meeting, Ann Arbor, Michigan, June 1994.
15 Chichilinisky, G. 1996. The economic value of the earth's resources. *Trends in Ecology and Evolution* 11:135–40; De Soto, H. 1993. The missing ingredient. In 'The future surveyed: 150 *Economist* years', *Economist* 11 September 1993, pp. 8–10.
16 Ostrom, E., Walker, J. and Gardner, R. 1992. Covenants without a sword: self-governance is possible. *American Political Science Review* 86:404–17. A similar conclusion – that communication was important in solving tragedies of the commons – was reached using a different game by Edney, J. J. and Harper, C. S. 1978. The effects of information in a resource management problem. A social trap analogy. *Human Ecology* 6:387–95.
17 Diamond, J. 1993. New Guineans and their natural world. In: *The Biophilia Hypothesis* (eds. Kellert, S. R. and Wilson, E. O.). Island Press, Washington, DC.
18 Jones, D. N., Dekker, R. W. R. J. and Roselaar, C. S. 1995. *The Megapodes*. Oxford University Press, Oxford.
19 See the chapters by Eric Alden-Smith and Richard Lee. 1988. In: *Hunters and Gatherers. Vol. 1: History, Evolution and Social Change* (eds. Ingold, T., Riches, D. and Woodburn, J.). Berg, Oxford.
20 Brubaker. *Property Rights in the Defence of Nature. op. cit.*
21 Cashdan, E. 1980. Egalitarianism among hunters and gatherers. *American Anthropologist* 82:116–20.

22 Carrier, J. G. and Carrier, A. H. 1983. Profitless property: marine ownership and access to wealth on Ponam Island, Manus Province. *Ethnology* 22:131–51.
23 Osborne, P. L. 1995. Biological and cultural diversity in Papua New Guinea: conservation, conflicts, constraints and compromise. *Ambio* 24:231–7.
24 Brubaker. *Property Rights in the Defence of Nature. op. cit.*; Anderson, T. (ed.) 1992. *Property Rights and Indian Economies*. Rowman and Littlefield, Lanham, Maryland.

CHAPTER THIRTEEN

1 Quoted in Webb, R. K. 1960. *Harriet Martineau: A Radical Victorian*. Heinemann, London.
2 Many attempts have been made to attribute the different societies of northern and southern Italians to genetic differences, but few are convincing. See Kohn, M. 1995. *The Race Gallery*. Jonathan Cape, London.
3 Putnam, R. 1993. *Making Democracy Work: Civil Traditions in Modern Italy*. Princeton University Press, Princeton. Fukuyama, F. 1995. *Trust: The Social Virtues and the Creation of Prosperity*. Hamish Hamilton, London.
4 Masters, R. D. 1996. *Machiavelli, Leonardo and the Science of Power*. University of Notre Dame Press, Indiana; Passmore, J. 1970. *The Perfectibility of Man*. Duckworth, London.
5 Hobbes, T. 1651/1973. *Leviathan*. Introduction by Kenneth Minogue. J. M. Dent and Sons, London.
6 Malthus, T. R. 1798/1926. *An Essay on the Principle of Population as it affects the future Improvement of Society, with Remarks on the Speculations of Mr Godwin, M. Condorcet and other Writers*. Facsimile edition, Macmillan, London. See also Ghiselin, M. T. 1995. Darwin, progress, and economic principles. *Evolution* 49:1029–37. I believe, eccentrically, that Darwin exaggerated his debt to Malthus in order to disguise his debt to Erasmus Darwin, his scandalous grandfather. Erasmus's great poem *The Temple of Nature*, published posthumously in 1801, was heavily influenced by Malthus, and Darwin certainly read it long before he read Malthus in 1828. Desmond King-Hele, Erasmus's biographer, also thinks this.
7 Jones, L. B. 1986. The institutionalists and *On the Origin of Species*: a case of mistaken identity. *Southern Economic Journal* 52:1043–55; Gordon, S. 1989. Darwin and political economy: the connection reconsidered. *Journal of the History of Biology* 22: 437–59.

8 Huxley, T. H. 1888. The struggle for existence in human society. *Collected Essays 9*.
9 The information on Hitler's eugenic sources, and the quotation from Wells, are from Watson, G. 1985. *The Idea of Liberalism*. Macmillan, London.
10 Degler, C. 1991. *In Search of Human Nature: The Decline and Revival of Darwinism in American Social Thought*. Oxford University Press, New York.
11 Rousseau, J.-J. 1755/1984. *A Discourse on Inequality*. Penguin, Harmondsworth.
12 Quoted in Graham, H. G. 1882. *Rousseau*. William Blackwood and Sons, Edinburgh.
13 Wreckage of La Pérouse's two vessels, *L'Astrolabe* and *La Boussole*, was found twenty-eight years later off Vanikoro island, north of the New Hebrides. His account of the voyage was published posthumously from notes he sent to Paris in 1787.
14 Moorehead, A. 1966. *The Fatal Impact: An Account of the Invasion of the South Pacific, 1767–1840*. Hamish Hamilton, London; Neville-Sington, P. and Sington, D. 1993. *Paradise Dreamed*. Bloomsbury, London.
15 Freeman, D. 1995. The debate, at heart, is about evolution. In: *The Certainty of Doubt: Tributes to Peter Munz* (eds. Fairburn, M. and Oliver, W. H.). Victoria University Press, Wellington, New Zealand.
16 Freeman, D. 1991. Paradigms in collision. A public lecture given at the Australian National University, 23 October 1991; Freeman, D. 1983. *Margaret Mead and Samoa: The Making and Unmaking of an Anthropological Myth*. Harvard University Press, Cambridge, Mass.; Wright, R. 1994 *The Moral Animal*. Pantheon, New York.
17 Quoted in Passmore. *The Perfectibility of Man. op. cit.*
18 See Robert Wright, in the *New Republic*, 28 November 1994, p. 34.
19 Chang, Jung. 1991. *Wild Swans: Three Daughters of China*. HarperCollins, London. See also Wright. *The Moral Animal. op. cit.*
20 Simon, H. 1990. A mechanism for social selection and successful altruism. *Science* 250:1665–8.
21 Fox, R. 1989. *The Search for Society: Quest for a Biosocial Science and Morality*. Rutgers University Press, New Brunswick.
22 Hazlitt, W. 1902. A reply to the essay on population by the Rev. T. R. Malthus. *The Collected Works of William Hazlitt. Vol. 4*. J. M. Dent, London.
23 Stewart, J. B. 1992. *Den of Thieves*. Touchstone, New York.
24 Wright. *The Moral Animal. op. cit.*.

25 Hayek, F. A. 1979. *Law, Legislation and Liberty. Vol. 3: The Political Order of a Free People*. University of Chicago Press, Chicago.
26 *Time*, 25 December 1995.
27 Duncan, A. and Hobson, D. 1995. *Saturn's Children*. Sinclair-Stevenson, London.

Index